825 -

E. Petroleum —
1st 7td 1896

vg— owner stamp on
front f—ep. 1st ptd
1896

377

384748204
3A-3

GAS AND FUEL ANALYSIS
FOR ENGINEERS.

A COMPEND FOR THOSE INTERESTED IN THE ECONOMICAL APPLICATION OF FUEL.

PREPARED ESPECIALLY FOR THE USE OF STUDENTS
at the
MASSACHUSETTS INSTITUTE OF TECHNOLOGY.

BY

AUGUSTUS H. GILL, S.B., Ph.D.,

Associate Professor of Technical Analysis at the
Massachusetts Institute of Technology, Boston, Mass.
Author of "A Short Handbook of Oil Analysis,"
"Engine Room Chemistry."

FIFTH EDITION, REVISED.

FIRST THOUSAND.

NEW YORK:
JOHN WILEY & SONS.
London: CHAPMAN & HALL, Limited.
1908.

The Scientific Press
Robert Drummond and Company
New York

PREFACE.

———

THIS little book is an attempt to present in a concise yet clear form the methods of gas and fuel analysis involved in testing the efficiency of a boiler plant. Its substance was given originally, in the form of lectures and heliotyped notes, to the students in the courses of Chemical, Mechanical, and Electrical Engineering, but in response to requests it has been deemed expedient to give it a wider circulation.

At the time of its conception, nothing of the kind was known to exist in the English language; in German we now have the excellent little book of Dr. Ferdinand Fischer, "Taschenbuch für Feuerungs-Techniker."

The present book is the result of six years' experience in the instruction of classes of about one hundred students. It is in no sense a copy of any other work, nor is it a mere compilation. The author has in every case endeavored to give credit where anything has been taken from outside sources; it is, however, difficult to credit single ideas, and if he has been remiss in this respect it has been unintentional.

The study of flue-gas analysis enables the engineer to investigate the various sources of loss; and if this compend stimulates and renders easy such investigation, the writer's purpose will have been accomplished. The necessary apparatus can be obtained from the leading dealers in New York City.

The author wishes to acknowledge his indebtedness to our former Professor of Analytical Chemistry, Dr. Thomas M. Drown, and to Mrs. Ellen H. Richards, by whose efforts the department of Gas Analysis was established.

He will also be grateful for any suggestions or corrections from the profession.

MASSACHUSETTS INSTITUTE OF TECHNOLOGY,
BOSTON, November, 1896.

PREFACE TO THE FIFTH EDITION.

THE changes made in the present edition include a brief treatment of the subjects of the Storage of Coal and its Spontaneous Combustion, and, in view of the increasing use of liquid fuel, methods of its analysis and testing.

As in the past, minor additions and corrections have been made where necessary to bring the book up to present practice.

MASSACHUSETTS INSTITUTE OF TECHNOLOGY,
BOSTON, August, 1908.

CONTENTS.

CONTENTS.

CHAPTER VII.

CHAPTER VIII.

APPENDIX.

LIST OF ILLUSTRATIONS.

GAS AND FUEL ANALYSIS.

CHAPTER I.

INTRODUCTION AND METHODS OF SAMPLING.

UNTIL within recent years, the mechanical engineer in testing a boiler plant has been compelled to content himself with the bare statement of its efficiency, little or no idea being obtained as to the apportionment of the losses. Knowing the composition and temperature of the chimney-gases and the analysis of the coal and ash, the loss due to the formation of carbonic oxide, to the imperfect combustion of the coal, to the high temperature of the escaping gases, can each be determined and thus a basis for their reduction to a minimum established.

By the simple analysis of the chimney-gases and determination of their temperature, a very good idea of the efficiency of the plant can be obtained previous to making the engineering test. For example, in a test which the author made in connection with his students, the efficiency was increased from 58 to 70 per cent, upon the results of the gas analysis alone.

To this end a representative sample must be collected according to the method about to be described.

SAMPLING.

Before proceeding to take a sample of the gas, the plant—for example, a boiler setting—from which the gas is to be taken should be thoroughly inspected, and all apertures by which the air can enter, carefully stopped up. A suitable tube is then inserted air-tight in the gas-duct, connected with the sampling or gas apparatus, and suction applied, thus drawing the gas out. Cork, putty, plaster of Paris, wet cotton-waste, or asbestos may be used to render the joint gas-tight. The place of insertion should be chosen where the gas will be most completely mixed and least contaminated with air. The oil-bath containing the thermometer is similarly inserted near the gas-tube, and the temperature read from time to time.

1. **Tubes.**—The tubes usually employed are Bohemian-glass combustion tubing or water-cooled metal tubes; those of porcelain or platinum are also sometimes used. Glass and porcelain tubes when subjected to high temperatures must be previously warmed or gradually inserted: the former may be used up to temperatures of 600° C. (1200° F.). Uncooled metal tubes, other than those of platinum, should under no circumstances be used.*

---

* Fischer, "Technologie der Brennstoffe," 1880, p. 221, states that the composition of a gaseous mixture was changed from 1.5 to 26.0 per cent carbon dioxide, by the passage through an iron tube heated to a dull red heat, the carbonic oxide originally present reducing the iron oxide with the formation of carbon dioxide.

The metal tube with the water cooling is made as shown in Fig. 1, *c* being a piece of brass pipe 3 feet long, $1\frac{1}{4}$ inches outside diameter, *b* the same length, $\frac{5}{8}$ inch in diameter, and *a* $\frac{1}{4}$ inch in diameter. The water enters at *d* and leaves at *e*. The walls of

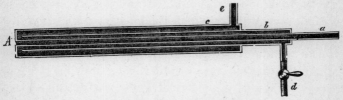

FIG. 1.—GAS-SAMPLING TUBE.

the tubes are $\frac{1}{16}$ inch thick. The joint at *A* should be brazed; the others may be soldered.

Platinum tubes from their high cost and small bore are seldom used; they are attacked by carbon, cyanogen, arsenic, and metallic vapors.

2. Apparatus for the Collection of Samples.— A convenient sampling apparatus is shown in Fig. 2. It may be made from a liter separatory funnel—instead of the bulb there shown—fitted with a rubber stopper carrying a tube passing to the bottom and a T tube; both of these, except where sulphur-containing gases are present, can advantageously be made of $\frac{3}{16}$-inch lead pipe. The stopper should not be fastened down with wire between the tubes after the manner of wiring effervescent drinks, as this draws the rubber away from the tubes and occasions a leak. The fastening shown consists of a brass plate fitting upon the top of the stopper, provided with screws and nuts which pass through a wire around the neck of the separatory. A chain fastened to the plate serves as a convenient method of handling it.

In using the apparatus, the bulb is filled with water by connecting the stem with the water-supply and opening one of the pinchcocks upon the T tube; the

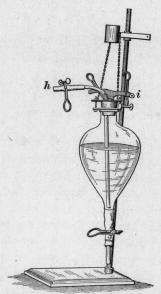

FIG. 2.—SAMPLING APPARATUS.

water thus entering from the bottom forces the air out before it. One branch of the T is connected with the sampling-tube and the other with the suction-pump, the stopcocks being open, and a current of gas drawn down into the pump; upon opening the cock upon the stem, the water runs out, drawing a small portion of the gas-current passing through the T after it into the bulb. It is then taken to a convenient

place for analysis, the tube *h* connected with a head of
water, a branch of the T *i*, with the gas apparatus, and
a sample of gas forced over into the letter for analysis.

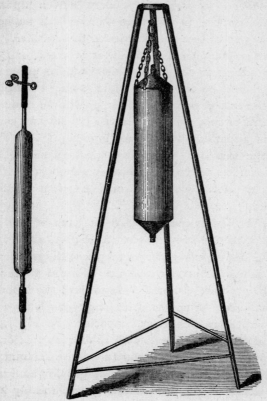

FIG. 4.—GAS-TUBE.   FIG. 3.—SAMPLING APPARATUS FOR
MINE-GASES.

Enough water should be left in the bulb to seal the
stopcock on the bottom and prevent leakage.   This
apparatus is better adapted for the needs of the class-

room than for actual practice, as it enables the same sample to be given to eight or ten students. As has been shown by several years' experience, the water exercises no appreciable solvent action upon the gaseous mixture in the time—about half an hour—necessary to collect and distribute the samples. It is often necessary to attach about a yard of $\frac{1}{4}$-inch rubber tubing to the stem of the bulb to prevent air being sucked up through it when taking a sample.

In the actual boiler-test it is preferable to insert a T instead of this apparatus in the gas-stream, connect the gas apparatus to the free branch of this T, and draw the sample. In making connections with gas apparatus the air in the rubber connectors should be displaced with water by means of a medicine-dropper.

In the Saxon coal-mines, zinc cans of ten liters capacity, of the form shown in Fig. 3, are used by Winkler for sampling the mine-gases; they are carried down filled with water and this allowed to run out, and the gas thus obtained brought into the laboratory and analyzed. Small samples of gas may very well be taken in tubes of 100 cc. capacity like Fig. 4, the ends of which are closed with rubber connectors and glass plugs. Rubber bags are not to be recommended for the collection and storage of gas for analysis, as they permit of the diffusion of gases, notably hydrogen.

3. **Apparatus for Producing Suction.**—I. WATER-PUMPS—(*a*) Jet-pumps, depending for their action upon a considerable head of water, and (*b*) those depending rather upon a sufficient fall of water.

(*a*) *Jet-pumps.*—The Richards' jet pump * is shown in section in Fig. 5 and much resembles a boiler injector; it consists of a water-jet *w*, a constriction or waist *a*, a waste-tube *o*, and a tube for the inspiration of air. The jet of water forms successive pistons across *a*, drawing the air in with it and is broken up into foam by the zigzag tube *o*.

This pump is known in Germany as Muencke's, and in England as Wing's; Chapman's pump is also a modified form.

It may be easily constructed in glass, the jets passing through rubber stoppers which are wired down, thus admitting of adjustment to the conditions under which it has to work.†

(*b*) *Fall-pumps.*—Bunsen's pump, Fig. 6, consists of a wide glass tube *A*, drawn out at the bottom for connection with a $\frac{1}{4}$-inch lead pipe *b*, and at the top for connection with *c*, the tube through which the air is drawn; this tube is usually fused in, although it may be connected with rubber; *a* is a rubber tube provided with screw cocks connected with the water-supply; *d* is connected with a mercury column, and the vessel *B* serves for the retention of any water which might be drawn back into the apparatus evacuated.

The tube *b* for the best results should be 32 feet in length, equal to the height of a column of water supported by the atmosphere, although for the ordinary purposes of gas-sampling it may be shorter.

When water is admitted through *a* it fills *b*, acting

---

* Richards, Am. Jour. of Science (3), **8**, 412; Trans. Am. Inst. Min. Engrs., **6**, 492 (1874).

† The pump will also work well using steam.

as a continually falling piston drawing the current of air through *e* and its connections. These various forms of water-pumps should give a vacuum repre-

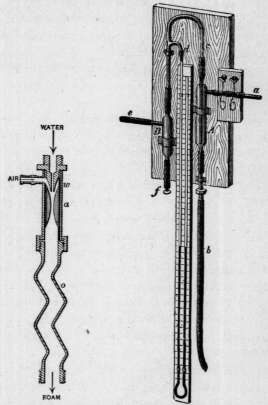

FIG. 5.—RICHARDS' JET-PUMP.   FIG. 6.—BUNSEN'S PUMP.

sented by the height of the barometer less the tension of aqueous vapor at the temperature at which they are used, or about 29 inches of mercury.

II. STEAM-PUMPS.—Kochinke describes the apparatus in use in the Muldner Hütten in Freiberg, shown at one-fifth size in Fig. 7. It consists of a glass tube drawn down to an opening 6 mm. in diam-

FIG. 7.—STEAM AIR-PUMP.

eter; concentric with this, and held in place by the washer a, is the steam-jet 2 mm. in diameter, passing through the cork b, the cement c, and covering d. It is connected with the steam-pipe at g by webbed rubber tubing f; the air enters at e. This is said to give very good results and be economical in use of steam.

In case neither water nor steam be available, recourse must be had to the ordinary rubber syringe-bulbs, provided with suitable valves, obtainable at any rubber store, or to a bottle aspirator. This consists of two one-gallon bottles, provided with doubly perforated rubber stoppers, carrying tubes of glass or lead bent at right angles. In each bottle one of these tubes passes nearly to the bottom, and these are connected together by a piece of rubber tubing a yard long, carrying a screw pinchcock. The other tube in each case stops immediately under the stopper. Upon filling one of the bottles with water, inserting the stopper and blowing strongly through the short tube, water will fill the long tubes thus forming a siphon, and upon lowering the empty bottle, a current of air will be sucked in through the short tube originally blown into; this may be regulated by the screw pinchcock.

In inserting the gas-sampling tube care should be taken not to insert it so close to the source of heat as to draw out the gases in a dissociated, i.e. partly decomposed, condition.

In case of very smoky fuels it is well to filter the gases through rolls of fine wire gauze or asbestos; in sucking them through a washing-bottle, the water may change the composition of the sample.

CHAPTER II.

APPARATUS FOR THE ANALYSIS OF CHIMNEY-GASES.

IN the writer's opinion the apparatus which is best adapted for this purpose is that of Orsat; it is readily portable, not liable to be broken, easy to manipulate, sufficiently accurate, and—in the modification about to be described—always ready for use, there being no stopcocks to stick fast.

As the Bunte and Elliott apparatus are also used for this purpose, they too will be described.

Fischer's apparatus, using mercury, is rather too difficult for the average engineer; Hempel's or Morehead's * apparatus for the analysis of illuminating-gas might also be used; it is, however, not customary.

ORSAT APPARATUS.

Description.—The apparatus Fig. 8, is enclosed in a case to permit of transportation from place to place; furthermore, the measuring-tube is jacketed with water to prevent changes of temperature affecting the gas-volume. The apparatus consists essentially of the levelling-bottle A, the burette B, the pipettes P', P'', P''', and the connecting tube T.

---

* No. 143 Lake Street, Chicago.

Manipulation.— The reagents in the pipettes should be adjusted in the capillary tubes to a point on the stem about midway between the top of the pipette and the rubber connector. This is effected by opening wide the pinchcock upon the connector, the

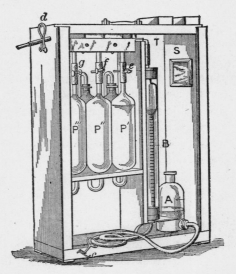

FIG. 8.—ORSAT'S GAS APPARATUS.

bottle being on the table, and very gradually lowering the bottle until the reagent is brought to the point above indicated. Six inches of the tubing used correspond to but 0.1 cc., so that an error of half an inch in adjustment of the reagent is without influence upon the accuracy of the result. The reagents having been thus adjusted, the burette and connecting tube are completely filled with water by opening *d* and raising the levelling-bottle. The apparatus is now ready to receive a sample of gas (or air for practice). In case a flue-gas is to be analyzed d is connected with i, Fig. 2, A lowered and about 102 cc. of the gas forced over by opening h; or d may be connected with a T joint in the gas-stream; the burette after filling is allowed to drain one minute by the sand-glass, c snapped upon its rubber tube, and the bottle A raised to the top of the apparatus. By gradually opening c the water is allowed to run into the burette until the lower meniscus stands upon the 100 or 0 mark (according to the graduation of the apparatus). The gas taken is thus compressed into the space occupied by 100 cc., and by opening d the excess escapes. Open c and *bring the level of the water in the bottle to the same level as the water in the burette* and take the reading, which should be 100 cc. Special attention is called to this method of reading: if the bottle be raised, the gas is compressed; if lowered, it is expanded.

Determination of Carbon Dioxide.—The gas to be analyzed is invariably passed first into pipette P', containing potassium hydrate for the absorption of carbon dioxide, by opening e and raising A. The gas displaces the reagent in the front part of the pipette, laying bare the tubes contained in it, which being covered with the reagent present a large absorptive surface to the gas; the reagent moves into the rear arm of the pipette, displacing the air over it into the flexible rubber bag which prevents its diffusion into the air. The gas is forced in and out of the pipette by raising and lowering A, the reagent finally brought approximately to its initial point on the stem of the pipette, the burette allowed to drain one minute, and the reading taken. The difference between this and the initial reading represents the cubic centimeters of carbon dioxide present in the gas. To be certain that all the carbon dioxide is removed, the gas should be passed a second time into P' and the reading taken as before; these readings should agree within 0.1 per cent.

Determination of Oxygen.—The residue from the absorption of carbon dioxide is passed into the second pipette, P'', containing an alkaline solution of potassium pyrogallate, until no further absorption will take place. The difference between the reading obtained and that after the absorption of carbon dioxide, represents the number of cubic centimeters of oxygen present.

Determination of Carbonic Oxide.—The residue from the absorption of oxygen is passed into the third pipette, P''', containing cuprous chloride, until no further absorption takes place; that is, in this case until readings agreeing exactly (not merely to 0.1) are obtained. The difference between the reading thus obtained and that after the absorption of oxygen, represents the number of cubic centimeters of carbonic oxide present.

Determination of Hydrocarbons. — The residue left after all absorptions have been made may consist, in addition to nitrogen, the principal constituent, of hydrocarbons and hydrogen. Their determination is difficult for the inexperienced, and, if desired, a sample of the flue-gas should be taken, leaving as little water in the apparatus as possible, and sent to a competent chemist for analysis.

Accuracy.—The apparatus gives results accurate to 0.2 of one per cent.

Time Required.—About twenty minutes are required for an analysis; two may be made in twenty five minutes, using two apparatus.

Notes.—The method of adjusting the reagents is the only one which has been found satisfactory: if the bottle be placed at a lower level and an attempt made to shut the pinchcock c upon the connector at the proper time, it will almost invariably result in failure.

The process of obtaining 100 cc. of gas is exactly analagous to filling a measure heaping full of grain and striking off the excess with a straight-edge; it saves arithmetical work, as cubic centimeters read off represent percent directly.

It often happens when e is opened, c being closed, that the reagent in P' drops, due not to a leak as is usually supposed, but to the weight of the column of the reagent expanding the gas.

The object of the rubber bags is to prevent the access of air to the reagents, those in P'' and P''' absorbing oxygen with great avidity, and hence if freely exposed to the air would soon become useless.

Carbon dioxide is always the first gas to be removed from a gaseous mixture. In the case of air the percentage present is so small, 0.08 to 0.1, as scarcely to be seen with this apparatus. It is important to use the reagents in the order given; if by mistake the gas be passed into the second pipette, it will absorb not only oxygen, for which it is intended, but also carbon dioxide; similarly if the gas be passed into the third pipette, it will absorb not only carbonic oxide, but also oxygen as well.

The use of pinchcocks and rubber tubes, original with the author, although recommended by Naef,* is considered by Fischer,† to be inaccurate. The experience of the author, however, does not support this assertion, as they have been found to be fully as accurate as glass stopcocks, and very much less troublesome and expensive.

In case any potassium hydrate or pyrogallate be sucked over into the tube T or water in A, the analysis is not spoiled, but may be proceeded with by connecting on water at d, opening this cock, and allowing the water to wash the tubes out thoroughly. The addition of a little hydrochloric acid to the water in the bottle A will neutralize the hydrate or pyrogallate, and the washing may be postponed until convenient.

After each analysis the number of cubic centimeters of oxygen and carbonic oxide should be set down upon the ground-glass slip provided for the purpose. By adding these numbers and subtracting their sum from the absorption capacity (see Reagents) of each reagent, the condition of the apparatus is known at any time, and the reagent can be renewed in season to prevent incorrect analyses.

BUNTE APPARATUS.

Description.—The apparatus Fig. 9 consists of a burette—bulbed to avoid extreme length—provided

---

* Wagner's Jahresb. 1885, p. 423.
† Technologie d. Brennstoffe, foot note p. 295.

at the top with a funnel *F* and three-way cock *j*, and a cock *l* at the bottom. These stopcocks are best of the Greiner and Friedrichs obliquely bored form. The burette is supported upon a retort-stand with a spring clamp.

A "suction-bottle" *S*, an 8-oz. wide-mouthed bottle, fitted similarly to a wash-bottle, except that the delivery-tube is straight and is fitted with a four-inch piece of $\frac{1}{4}$-inch black rubber tubing, serves to withdraw the re-agents and water when necessary. A reservoir to contain water at the tem-perature of the room, fitted with a long rubber tube, should be provided for washing out the reagents and filling the burette.

Manipulation. — Before using the apparatus, the keys of the stopcocks should be taken out, wiped dry, to-gether with their seats, and sparingly smeared with vaseline or a mixture of vaseline and tallow and replaced. The completeness of the lubrication can be judged by the transparency of the joint, a thoroughly lubricated joint showing no ground glass. The burette is filled with water by attaching the rubber tube to the tip at *l* and opening the stopcocks at the top and bottom; *j* is connected with the source whence the gas is to be taken, turned to communicate with the burette and opened, about 102 cc. of gas allowed to run in, and *j* and *l* closed.

FIG. 9.—BUNTE'S GAS APPARATUS.

The cup F is filled with water to the 25-cc. mark, j turned to establish communication between it and the burette, the burette allowed to drain one minute by the sand-glass, and the reading taken, the cup being refilled to the mark if necessary. The readings are thus taken under the same pressure each time, i.e., this column of water plus the height of the barometer; and as the latter is practically constant during the analysis, no correction need be applied, it being within the limits of error.

Determination of Carbon Dioxide.—The " suction-bottle " is connected with the tip of the burette, l opened, and the water carefully sucked out nearly to l. The bottle is now disconnected, the burette dismounted from its clamp, using the cup as a handle, and the 25 cc. of water turned out. The tip is immersed under potassium hydrate contained in the No. 3 porcelain dish, and the cock l opened, then closed, and the tip wiped clean with a piece of cloth. The burette is now shaken, holding it by the tip and the cup, the thumbs resting upon j and l; more reagent is introduced, the absorption of the gas causing a diminished pressure, and the operation repeated until no change takes place. The cup is now filled with water, j opened, and the reagent completely washed out into an ordinary tumbler placed beneath the burette. Four times filling of F should be sufficient for this purpose. The cup is now filled to the 25-cc. mark, j opened, and the reading taken as before.

The difference between this reading and the initial represents the number of cubic centimeters of carbon dioxide; this divided by the volume of the gas taken gives the per cent of this constituent.

Determination of Oxygen.—The water is again sucked out, and potassium pyrogallate solution introduced, similarly to potassium hydrate; this is displaced by water, and the reading taken as before. The difference between this and the last reading is the volume of oxygen present.

Determination of Carbonic Oxide.—The water is removed for a third time, and acid cuprous chloride solution introduced and the absorption made as before; this is washed out, first with water containing a little hydrochloric acid to dissolve the white cuprous chloride which is precipitated by the addition of water, and finally with pure water, and the reading taken as before. The difference between this and the preceding gives the volume of carbonic oxide present.

Notes.—Especial care should be taken not to grasp the burette by the bulb, as this warms the gas and renders the readings inaccurate. The stopcocks can conveniently be kept in the burette by elastic bands of suitable size. When the apparatus is put away for any considerable time, a piece of paper should be inserted between the key and socket of each stopcock to prevent the former from sticking fast. To ascertain when the absorption is complete, the burette is mounted in its clamp and allowed to drain until the meniscus is stationary, the dish containing the reagent raised until the tip is covered, *l* opened, and any change in level noted. If the meniscus rises, the absorption is incomplete and must be continued; if it remains stationary or falls, the absorption may be regarded as finished. In case the grease from the stopcocks becomes troublesome inside the burette, it may be removed by dissolving it in chloroform and washing out with alcohol and then with water. The object in sucking the water not quite down to *l*, thus leaving a little water in the burette, is to discover if *l* leaks, the air rushing in causes bubbles.

The object in washing out each reagent and taking all readings over water is to obviate corrections for the tension of aqueous vapor over potassium hydrate, hydrochloric acid, or any of the reagents which might be employed. The tension of aqueous vapor over seven per cent caustic soda is less than over water.

Accuracy and Time Required.—The apparatus is rather difficult to manipulate, but fairly rapid—about twenty-five minutes being required for an analysis—and accurate to one tenth of one per cent.

ELLIOTT APPARATUS.

Description.—The apparatus Fig. 10 consists of a burette holding 100 cc. graduated in tenths of a cubic centimeter and bulbed like the Bunte apparatus—the bulb holding about 30 cc.; it is connected with a levelling-bottle similar to the Orsat apparatus. The top of the burette ends in a capillary stopcock, the stem of which is ground square to admit of close connection with the "laboratory vessel," an ungraduated tube similar to the burette, except of 125 cc. capacity. The top of this "vessel" is also closed with a capillary stopcock, carrying by a ground-glass joint a thistle-tube *F*, for the introduction of the reagents. The lower end of this "vessel" is closed by a rubber

stopper carrying a three-way cock *o*, and connected
with a levelling-bottle *D*. The
burette and vessel are held upon a
block of wood—supported by a ring
stand—by fine copper wire tightened by violin keys.

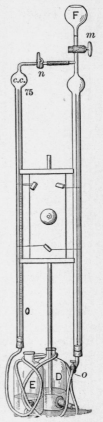

Manipulation.—The ground-glass
joints are lubricated as in the Bunte
apparatus. The levelling-bottles are
filled with water, the stopcocks
opened, and the bottles raised until
the water flows through the stopcocks *m* and *n*. *m* is connected
with the source whence the gas to
be analyzed is to be taken, *n* closed,
D lowered and rather more than 100
cc. drawn in, and *m* closed. *n* is
opened, *D* raised and *E* lowered,
nearly 100 cc. of gas introduced,
and *n* closed; by opening *m* and
raising *D* the remainder of the gas
is allowed to escape, the tubes being
filled with water and *m* closed. *n* is
opened and the water brought to
the reference-mark; the burette is
allowed to drain one minute, **the**
level of the water in *E* is brought
to the same level as in the burette,
and the reading taken.

Fig. 10.—Elliott
Gas Apparatus.

Determination of Carbon Dioxide.—By raising *E*,
opening *n*, and lowering *D*, the gas is passed over into
the laboratory vessel; *F* is filled within half an inch

of the top with potassium hydrate, o closed, m opened, and the reagent allowed to slowly trickle in. A No. 3 evaporating-dish is placed under o, and this turned to allow the liquid in the laboratory vessel to run into the dish. At first this is mainly water, and may be thrown away; later it becomes diluted reagent and may be returned to the thistle-tube. When the depth of the reagent in the thistle-tube has lowered to half an inch, it should be refilled either with fresh or the diluted reagent and allowed to run in until the absorption is judged to be complete, and the gas passed back into the burette for measurement. To this end close o and then m, raise E, open n, and force some pure water into the laboratory vessel, thus rinsing out the capillary tube. Now raise D and lower E, shutting n when the liquid has arrived at the reference-mark. The burette is allowed to drain a minute, the level of the water in the bottle E brought to the same level as the water in the burette, and the reading taken.

Determination of Oxygen.—The manipulation is the same as in the preceding determination, potassium pyrogallate being substituted for potassium hydrate; the apparatus requiring no washing out.

Determination of Carbonic Oxide.—The laboratory vessel, thistle-tube, and bottle if necessary, are washed free from potassium pyrogallate and the absorption made with acid cuprous chloride similarly to the determination of carbon dioxide. The white precipitate of cuprous chloride may be dissolved by hydrochloric acid.

Accuracy and Time Required.—The apparatus is as accurate for absorptions as that of Orsat; it is stated to be much more rapid—a claim which the writer cannot substantiate. It is not as portable, is more fragile, and more troublesome to manipulate, and as the burette is not jacketed it is liable to be affected by changes of temperature.

Notes.—In case at any time it is desired to stop the influx of reagent, *o* should be closed first and then *m*; the reason being that the absorption may be so rapid as to suck air in through *o*, *m* being closed.

The stopcock should be so adjusted as to cause the reagent to spread itself as completely as possible over the sides of the burette.

By the addition of an explosion-tube it is used for the analysis of illuminating-gas,[*] bromine being used to absorb the "illuminants." Winkler [†] states that this absorption is incomplete; later work by Treadwell and Stokes, and also Korbuly,[‡] has shown that bromine water, by a purely physical solution, does absorb the "illuminants" completely; Hempel [§] states that explosions of hydrocarbons made over water are inaccurate, so that the apparatus can be depended upon to give results upon methane and hydrogen only within about two per cent.

---

[*] Mackintosh, Am. Chem. Jour. **9**, 294.
[†] Zeit. f. Anal. Chem. **28**, 286.
[‡] Treadwell's Quan. Analysis (Hall's translation), p. 569.
[§] Gasanalytische Methoden, p. 102.

GAS-BALANCES.

Under this heading are included various devices either for weighing a volume of the chimney-gas, as the Econometer of Arndt,[*] or weighing a globe in an atmosphere of the gas, as the Gas-balance of Custodis.[†] The Gas-composimeter of Uehling[‡] depends upon the laws governing the flow of gases through small apertures.

They are difficult to adjust and keep in adjustment, requiring to be checked frequently by the Orsat apparatus, and are expensive. Their indications are within about half of one per cent of those given by the chemical apparatus. Only the presence of carbon dioxide is indicated by them.

---

[*] Zeit. d. Vereins deutsch. Ingenieure, **37**, 801.

[†] Gill, Engine Room Chemistry, pp. 96 and 97.

[‡] Poole, Calorific Power of Fuels, p. 150, 2d edition (1900).

CHAPTER III.

MEASUREMENT OF TEMPERATURE.

In the majority of cases, the ordinary mercurial thermometer will serve to determine the temperature of the chimney-gases. It should not be inserted naked into the flue, but be protected by a bath of cylinder, or raw linseed oil, contained in a brass or iron tube. These tubes may be half an inch inside diameter and two to three feet in length. Temperatures as high as $625°$ C. have been observed in chimneys; this lasts of course but for a moment, but would be sufficient to burst the unprotected thermometer.

For the observation of higher temperatures, recourse must be had to the " high-temperature thermometers," filled with carbon dioxide under a pressure of about one hundred pounds, giving readings to $550°$ C.* These may be obtained of the dealers in chemical apparatus; some require no bath, being provided with a mercury-bath carefully contained in a steel tube, and the whole enclosed in a bronze tube.†

---

* Those made by W. Apel, Göttingen, Germany, are about three feet long, the scale occupying about one foot, thus avoiding the necessity of withdrawing the thermometer from the bath for reading. H. J. Green of Brooklyn, N. Y., makes similar ones.

† Hohmann Special Thermometers, made by Hohmann and Maurer Co., 42 High Street, Boston, Mass.

These thermometers should be tested from time to time either by comparison with a standard or by insertion in various baths of a definite temperature. Some of the substances used for these baths are: water, boiling-point 100°; naphthalene, Bpt. 219°; benzophenon, Bpt. 306°; and sulphur,* Bpt. 445°. Care should be taken that the bulb of the thermometer does not dip into the melted substance, but only into the vapor, and that the stem exposure be as nearly as possible that in actual use.

For the measurement of temperatures beyond the range of these thermometers the Le Chatelier thermo-electric pyrometer may be used. This consists of a couple formed by the junction of a platinum and platinum-10% rhodium wire, passing through fire-clay tubes in a porcelain or iron envelope and connected with a galvanometer. The hotter the junction is heated the greater the current and the galvanometer deflection; this latter is determined for several points, naphthalene, sulphur, and copper, Mpt. 1095° C., or even platinum, 1760° C., and a plot made with galvanometer-readings as abscissæ and temperatures as ordinates. From this the temperature corresponding to any deflection is readily obtained.

The exact description of the instrument and details of calibration are, however, beyond the scope of this work, and the student is referred for these to articles by Le Chatelier, Société Technique de l'Industrie du Gaz, 1890, abstracted in Jour. Soc. Chem. Industry.

---

* In testing the Hohmann thermometers in sulphur-vapor, the bronze tube should be prevented from corrosion by the vapor by a glass envelope.

9, 326, and Holman, Proc. Am. Academy, 1895, p. 234 ; later works are those of Le Chatelier and Boudouard, " High Temperature Measurements," transl. by G. K. Burgess (1901), and also C. L. Norton, " Notes on Heat Measurements " (1902).

An error of 5° in the reading of the thermometer affects the final result by about 20 calories.

In case neither of these methods be available nor

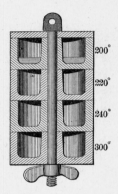

FIG. 11.—MELTING-POINT BOXES.

applicable, use may be made of the melting-points of certain metals or salts contained in small cast-iron boxes, Fig. 11. The melting-points of certain metals and salts are given in Table VII.

CHAPTER IV.

CALCULATIONS.

As has been already stated in the Introduction, the object of analyzing the flue-gases is to ascertain, first, the completeness of the combustion, especially the amount of air which has been used or the " pounds of air per pound of coal," and second, the amount of heat passing up chimney.

1. To Ascertain the Number of Pounds of Air per Pound of Coal.—A furnace-gas gives 11.5% CO_2, 7.4% O, 0.9% CO, and 80.2% N. Data: atomic weights, O = 16, C = 12; weight liter CO_2 = 1.966 grs., of O, 1.43 grs., of CO, 1.251 grs. Find the number of grams of each constituent in 100 liters of the furnace-gas, and from this the weight of carbon and weight of oxygen. 11.5 (liters CO_2) $\times$ 1.97 (wt. liter CO_2) = 22.66 grms. CO_2; now $\frac{32}{44}\left(\frac{O_2}{CO_2}\right)$ of this is oxygen = 16.48 grms., 6.18 grms. is carbon. The weight of free oxygen is $7.4 \times 1.43 = 10.58$ grms. The weight of carbon and oxygen in the carbonic oxide is $0.9 \times 1.25 = 1.12$ grms. CO. Now $\frac{16}{28}\left(\frac{O}{CO}\right)$ is oxygen or 0.64 grm., and 0.48 grm. is carbon. There are then present in 100 liters of the gas 27.70 grms. oxygen and 6.66 grms. carbon; corresponding to 120.0 grms. air to 6.66 grms. carbon, air being 23.1% oxygen by weight; or 18.02 $\left.\begin{array}{c}\text{grms.}\\\text{lbs.}\end{array}\right\}$ air per $\left.\begin{array}{c}\text{grm.}\\\text{lb.}\end{array}\right\}$ *carbon*. If the coal be 83% carbon, this figure must be diminished accordingly, giving in this case 14.95 lbs. air per lb. of *coal*. Theory requires 11.54 lbs. air per lb. of carbon, but in practice the best results are obtained by increasing this from 50% to 100%.*

2. **To Ascertain the Quantity of Heat Passing up Chimney**.—Determine the volume of gas generated from one kilo of coal when burned so as to produce the gas the analysis of which has just been made according to the directions given. The chemical analysis of the coal is as follows: moisture 1.5%, sulphur 1.2%, carbon 83%, hydrogen 2.5%, ash 11.4%, oxygen and nitrogen (by difference) 0.4%. Then there are in one kilo of coal 830 grms. carbon, of this suppose but 800 to be burned, the remaining 30 grms. going into the ash; of the 800 grms. 618/666 or 742 grms. produced carbon dioxide, and 48/666 or 58 grms. produced carbonic oxide. From 6.18 grms. carbon were produced 11.5 liters carbon dioxide in the problem in 1; hence 742 grms. would furnish 1381 liters. $6.18:742::11.5:y.$ $y = 1381.$ Similarly 58 grms. carbon would furnish 109.0 liters carbonic oxide. $0.48:58::0.90:z.$ $z = 109.0.$ The volume of oxygen can be found by the proportion 11.5 (% CO_2): 7.4 (% O):: 1381 : x. $x = 888$ liters. In the same manner the nitrogen is found to be 9631 liters. $11.5:80.2::1381:u.$ $u = 9631.$ One kilo of coal under these conditions furnishes 1.381 cu. meters

---

* Scheurer-Kestner, Jour. Soc. Chem. Industry, **7**, 616.

carbon dioxide, 0.109 c. m. carbonic oxide, 0.888
c. m. oxygen, and 9.631 c. m. nitrogen.

The quantity of heat carried off by each gas is its
rise of temperature $\times$ its weight $\times$ its specific heat.
The specific heats of the various gases are shown in
the table below, and for facility in calculation, a column
is given obtained by multiplying the weight by the
specific heat; multiplying the volumes obtained in the
previous calculation by the numbers in this column
and by the rise in temperature gives the number of
calories (C) that each gas carries away.

TABLE OF SPECIFIC HEATS OF VARIOUS GASES.*

Name of Gas.	Sp. Heat.	Wt. of Cu. M. Kg.	Sp. Heat $\times$ Wt. of Cu. M.	Log.
Carbon dioxide (10°–350°).	0.234	1.97	0.463	9.6656
" monoxide.........	0.245	1.26	0.308	9.4886
Oxygen..................	0.217	1.43	0.311	9.4928
Nitrogen.................	0.244	1.26	0.306	9.4857
Aqueous vapor...........	0.480	0.80	0.387	9.5877

In the test the average temperature of the escaping
gases was 275° C.; that of the air entering the grate
was 25° C., a rise of temperature of 250° C. As
shown by the wet-and-dry-bulb thermometer, the air
was 50 per cent saturated with moisture.

The calculation of the heat carried away is then for:

	Cu. M.					C.
Carbon dioxide...........	1.381	$\times$	250	$\times$	0.463 =	160.0
Carbonic oxide...........	0.109	$\times$	"	$\times$	0.308 =	8.4
Oxygen..................	0.888	$\times$	"	$\times$	0.311 =	69.1
Nitrogen.................	9.631	$\times$	"	$\times$	0.306 =	737.0
Total.............	12.009					974.5

* Fischer, Tech. d. Brennstoffe, p. 267.

There is, however, another gas passing up chimney of which we have taken no cognizance, namely, water-vapor; this comes from the moisture in the coal, from the combustion of hydrogen in the coal, and from the air entering the grate; its volume is calculated as follows:

The moisture in the coal as found by chemical analysis was $1.5\% = 0.015$ kg.; the hydrogen in the coal was $2.5\% = 0.025$ kg. The amount of water this forms when burned is nine times its weight, 0.025 kg. $\times 9 = 0.225$ kg. The moisture in the air entering the grate would be, if completely saturated, 22.9 grams per cubic meter, as shown by Table I; it was, how-ever, but 50% saturated. The quantity is then, the volume of air used per kilogram of coal $\times$ moisture contained in it, or $12.009 \times 22.9 \times 0.50 = 0.137$ kg. The weight of aqueous vapor passing up chimney per kilogram of coal is $0.015 + 0.225 + 0.137 = 0.377$ kg.; the quantity of heat that this carries off is 0.377 $\times 250 \times 0.480 = 45.2$ C. The total quantity of heat passing up chimney is then 1019.7 C. The heat of combustion of this coal as found by Mahler's calori-metric bomb was 7220 C.; hence the percentage of heat carried off is $1020/7220 = 14.1\%$.

The preceding calculations though correct are tedious, so much so, as to almost preclude their use for an hourly observation of the firing. They should be employed, however, in making the final calculation of a boiler-test, using the averages obtained.

Shields * has combined the operations in

---

* *Power*, 1908.

1. Pounds of Air per Pound of Coal (p. 28), and obtains the following formula:

Pounds of air per pound of coal

$$= 231 \frac{\text{Per cent. carbon in coal}}{\text{Per cent. } CO_2 + \text{per cent. CO.}}$$

Similarly, **Per cent. heat lost**

$$= \frac{\text{Per cent. carbon in coal}}{\text{Heating value of coal}} \times \frac{200 + \text{per cent. } CO_2}{\text{Per cent. } CO_2 + \text{per cent. CO}} \times$$

rise in temperature in °C.

The values found by this equation are 0.5 per cent. low, as no cognizance has been taken of the water vapor.

In rapid work the following formula will be found more applicable: Let o and n represent the percentages of oxygen and nitrogen found in the chimney-gas; then the ratio of the air actually used to that theoretically necessary is expressed by the formula

$$\frac{21}{21 - \left(\frac{79 \, o}{n.}\right)}.$$

Applying it in the case of the flue-gas given, it becomes

$$\frac{21}{21 - \left(\frac{79 \times 7.4}{80.2}\right)} = \frac{21}{13.7} = 1.533 \text{ ratio.}$$

Multiplying this by 11.54, the theoretical number of pounds of air per pound of carbon, we obtain 17.69 as against 18.02 on page 28.

Bunte * has given a shorter method for the deter-

---

* Jour. f. Gasbeleuchtung, **43**, 637 (1900); Abstr. Jour. Soc. Chem. Industry, **19**, 887.

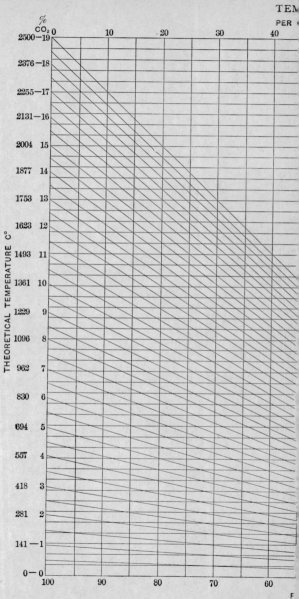

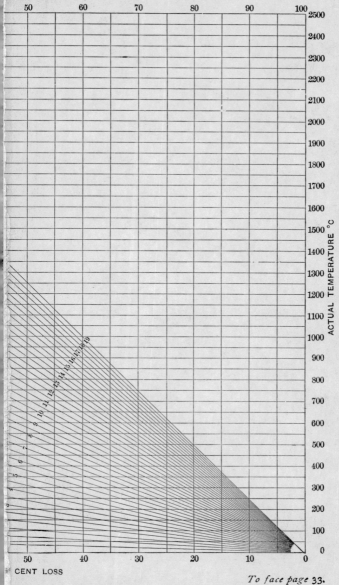

TABLE XL.

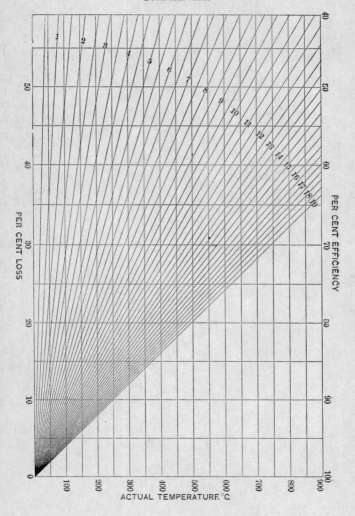

TABLE XI.

mination of the quantity of heat passing up chimney, and one which does not involve the analysis of the coal.

For every per cent of carbonic acid present 43.43 C. per cubic meter of flue-gases have been developed $= W$; $C =$ specific heat of the flue-gases per cubic meter; then W/C represents the initial temperature (which is never attained) the ratio of which to the actual exit temperature of the flue-gases shows the heat lost. If $T =$ this initial temperature and t the rise of temperature of the flue-gases, then t/T represents the heat lost in the chimney-gases.

The following table gives the data for the calculation for both pure carbon and coal of average value.

Per Cent of CO_2 in Chimney Gas.	Specific Heat of Chimney Gas.	Initial Temperature, W/C. Degrees C.		
		For Carbon $= T$.	For Coal $= T$.	Diff. for 0.1% CO_2.
1	0.308	141	167	
2	0.310	280	331	16
3	0.311	419	493	16
4	0.312	557	652	16
5	0.313	694	808	15
6	0.314	830	961	15
7	0.315	962	1112	15
8	0.316	1096	1261	15
9	0.318	1229	1407	15
10	0.319	1360	1550	14
11	0.320	1490	1692	14
12	0.322	1620	1830	14
13	0.323	1750	1968	14
14	0.324	1880	2102	13
15	0.324	2005	2237	13
16	0.325	2130	2366	13

Applying this to the problem on page 29 we find the initial temperature T to be 1762° C., the rise of temperature of the gases was 250° C., the loss is 250/1762 = 14.2%, against 14.1% found by the calculation page 31.

Bunte also employs a partially graphical method for the determination of the loss of heat. In Table X the extreme left-hand column represents the temperatures which should be obtained by the combustion of the average coal with the formation of a chimney-gas containing the percentages of carbon dioxide in the column next it. Applying this to our case we find the theoretical temperature for 11.5% CO_2 to be 1558°; dividing the rise of temperature actually observed— 250°—by this, we obtain 16.05%, or 2% more than by the method of page 31.

Almost the identical result can be obtained from Table XI directly: if the point of intersection of the diagonal representing the per cent of carbon dioxide with the horizontal line denoting the actual temperature, on the right, be followed to the bottom of the table the per cent of loss is ascertained.

Table XI is the lower right-hand corner of Table X enlarged.

W. A. Noyes[*] states that the following formula gives close results and is also independent of the composition of the coal.

$$\text{Percentage loss} = \left(0.011 + \frac{100 - \%CO_2}{\%CO_2}0.00605\right)(t' - t).$$

Lunge[†] has also given a shorter method for the calculation of the heat lost.

---

[*] Am. Chem. Journal, **19**, 162.
[†] Zeit. f. angewandte Chemie, 1889, 240.

The following table * shows roughly the excess of air and the per cent of heat lost in the chimney gases:

PER CENT OF CARBONIC ACID.

2	3	4	5	6	7	8	9	10	11	12	13	14	15

VOLUME OF AIR MORE THAN THEORY.

(Theory = 1.0).

9.5	6.3	4.7	3.8	3.2	2.7	2.4	2.1	1.9	1.7	1.6	1.5	1.4	1.3

PER CENT LOSS OF HEAT.

Temp. of chimney gases, 518° F.

90	60	45	36	30	26	23	20	18	16	15	14	13	12

Determination of Loss Due to Formation of Carbonic Oxide.—On page 29 we see that 58 grams of carbon burned to carbonic oxide; for every gram of carbon burned to carbonic oxide there is a loss of 5.66 C., in this case a loss of 328 C. The heating value of the coal is 7220 C., hence the loss is 4.5 per cent.

---

* Arndt's Econometer Circular.

CHAPTER V.

APPARATUS FOR THE ANALYSIS OF FUEL AND ILLUMINATING GASES.

HEMPEL'S APPARATUS.

Description.—The apparatus, Figs. 12 and 13, is very similar in principle to that of Orsat; the burette is longer, admitting of the reading of small quantities of gas, and the pipettes are separate and mounted in brass clamps on iron stands. P shows a "simple" pipette * provided with a rubber bag; this form, after ten years of use, can be said to satisfactorily take the place of the cumbersome "compound" pipette.

The pipette for fuming sulphuric acid † is shown at F, and differs from the ordinary in that vertical tubes after the manner of those in the Orsat pipettes replace the usual glass beads. This prevents the trapping of any gas by the filling, which was so common with the beads and glass wool. E represents the large explosion pipette, ‡ of about 250 cc. capacity, with walls half an inch thick; the explosion wires enter at the top and bottom to prevent short-circuiting; mercury is the confining liquid. The small explosion pipette holds

---

* Gill, Am. Chem. J., **14**, 231 (1892).

† *Id.*, J. Am. Chem. Soc., **18**, 67 (1896).

‡ Gill, J. Am. Chem. Soc., **17**, 771 (1895).

about 110 cc. and is of glass, the same thickness as
the simple pipettes. Water is here used as the confining liquid, and also usually in the burette.

An induction coil capable of giving a half-inch spark,

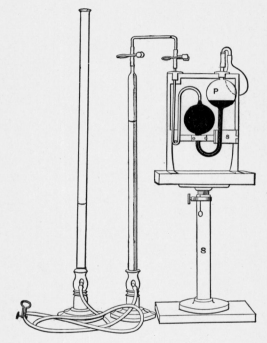

Fig. 12.—Showing Hempel Burette connected with the
Simple Pipette on the Stand.

with a six-cell "Samson" battery, four "simple"
pipettes and a mercury burette, complete the outfit.

The burette should be carefully calibrated and the
corrections may very well be etched upon it opposite
the 10-cc. divisions.

In working with the apparatus the pipettes are placed upon the adjustable stand *S* and connection made with the doubly bent capillary tube.

Manipulation.—To acquire facility with the use of the apparatus before proceeding to the analysis of

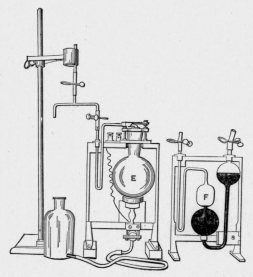

Fig. 13.—Explosion Pipette for Mercury and Sulphuric Acid Pipette.

illuminating-gas, it is well to make the following determinations, obtaining " check readings " in every case: I. Oxygen in air, by (1) absorption with phosphorus; (2) absorption with potassium (or sodium) pyrogallate; (3) by explosion with hydrogen.

I. DETERMINATION OF OXYGEN IN AIR.

(1) **By Phosphorus.**—100 cc. of air are measured out as with the Orsat apparatus, the burette being allowed to drain two minutes. The rubber connectors upon the burette and pipette are filled with water, the capillary tube inserted, as far as it will go, by a twisting motion, into the connector upon the burette, thus filling the capillary with water; the free end of the capillary is inserted into the pipette connector, the latter pinched so as to form a channel for the water contained in it to escape, and the capillary twisted and forced down to the pinch-cock. There should be as little free space as possible between the capillaries and the pinch-cock. Before using a pipette, its connector (and rubber bag) should be carefully examined for leaks, especially in the former, and if any found the faulty piece replaced.

The pinch-cocks on the burette and pipette are now opened, the air forced over into the phosphorus, and the pinch-cock on the pipette closed; action immediately ensues, shown by the white fumes; after allowing it to stand for fifteen minutes the residue is drawn back into the burette, the latter allowed to drain and the reading taken. The absorption goes on best at 20° C., not at all at below 15° C. ; it is very much retarded by small amounts of ethene and ammonia. No cognizance need be taken of the fog of oxides of phosphorus.

(2) **By Pyrogallate of Potassium.**—100 cc. of air are measured out as before, the carbon dioxide absorbed with potassium hydrate and the oxygen with potassium

pyrogallate, as with the Orsat apparatus; before setting aside the pyrogallate pipette, the number of cubic centimeters of oxygen absorbed should be noted upon the slate *s* on the stand.   This must never be omitted with any pipette save possibly that for potassium hydrate, as failure to do this may result in the ruin of an important analysis.   The reason for the omission in this case is found in the large absorption capacity—four to five litres of carbon dioxide—of the reagent.

(3) **By Explosion with Hydrogen.**—43 cc. of air and 57 cc. of hydrogen are measured out, passed into the small explosion pipette, the capillary of the pipette filled with water, the pinch-cocks and glass stop-cock all closed, a heavy glass or fine wire gauze screen placed between the pipette and the operator, the spark passed between the spark wires, and the contraction in volume noted.   *The screen should never be omitted, as serious accidents may occur thereby.*   The oxygen is represented by one third of the contraction.   For very accurate work the sum of the combustible gases should be but one sixth that of the non-combustible gases, otherwise some nitrogen will burn and high results will be obtained; * that is, $(H + O) : (N + H) :: 1 : 6$.

II. ANALYSIS OF ILLUMINATING-GAS.

100 cc. of gas are measured from the bottle containing the sample into the burette.

Determination of Carbon Dioxide.—The burette is connected with the pipette containing potassium

---

* This is shown in the work of Gill and Hunt, J. Am. Chem. Soc., **17**, 987 (1895).

hydrate and the gas passed into it with shaking until no further diminution in volume takes place.

Illuminants, C_nH_{2n}, C_nH_{2n-6} Series.—The rubber connectors are carefully dried out with filter-paper, a dry capillary used, and the gas passed into the pipette containing fuming sulphuric acid and allowed to stand, with occasional passes to and fro, for forty-five minutes. On account of the extremely corrosive nature of the absorbent it is not advisable to shake the pipette, as in case of breakage a serious accident might occur. For Boston gas this is sufficient, although with richer gases check readings to 0.2 cc. should be obtained. It is then passed into potassium hydrate, as in the previous determination, to remove any sulphurous acid which may have been formed and any sulphuric anhydride vapor, these having a higher vapor tension than water. The difference between this last reading and that after the absorption of the carbon dioxide represents the volume of "illuminants" or "heavy hydrocarbons" present.

As has already been stated, page 23, saturated bromine water may replace the fuming sulphuric acid. Fuming nitric acid is not recommended, as it is liable to oxidize carbonic oxide.

Oxygen.—This is absorbed, as in the analysis of air, by potassium or sodium pyrogallate.

Carbonic Oxide.—The gas is now passed into ammoniacal cuprous chloride, until the reading is constant to 0.2 cc.; it is then passed into a second pipette, which is fresh, and absorption continued until constant readings are obtained.

Gautier and Clausmann * have shown that some carbonic oxide escapes solution in cuprous chloride, so that for very accurate work it may be necessary to pass the gas through a U-tube containing iodic anhydride heated to 70° C.

This is done by interposing this tube between the burette and a simple pipette filled with potassium hydrate. The reaction is $5CO + I_2O_5 = 5CO_2 + 2I$. The diminution in volume represents directly the volume of carbonic oxide present.

The volume of air contained in the tube should be corrected for as follows: One end of the tube is plugged tightly and the other end connected with the gas burette partly filled with air. A bath of water at 9° C. is placed around the U-tube and the reading of the air in the gas burette recorded when constant; the bath is now heated to 100° and the burette reading again recorded when constant. The increase in reading represents one third the volume of the U-tube, $273 : 273 + (100 - 9) : : 3 : 4.$

Methane and Hydrogen.—(a) *Hinman's Method.*†
—The gas left from the absorption of carbonic oxide is passed into the large explosion pipette. About half the requisite quantity of oxygen (40 cc.) necessary to burn the gas is now added, mercury introduced through the T in the connector sufficient to seal the capillary of the explosion pipette, all rubber connectors carefully wired, the pinch-cocks closed, and the pipette cautiously shaken. A screen of heavy glass or fine wire gauze is interposed between the operator and the

---

* Bull. Soc. Chem. **35**, 513; Abstr. Analyst, **31**, 349 (1906).
† Gill and Hunt, J. Am. Chem. Soc., **17**, 987 (1895).

apparatus, the explosion wires are connected with the induction coil, a spark passed between them and the pinch-cocks opened, sucking in the remainder of the oxygen. The capillary is again sealed with mercury, the stop-cock opened and closed, to bring the contents of the pipette to atmospheric pressure, and the explosion repeated as before, and the stop-cock opened.

It may be found expedient, to increase the inflammability of the mixture, to introduce 5 cc. of "detonating-gas," the hydrolytic mixture of hydrogen and oxygen. The gas in the pipette containing carbon dioxide, oxygen, and nitrogen is transferred to the mercury burette and accurately measured. The carbon dioxide resulting from the combustion of the marsh-gas is determined by absorption in potassium hydrate; to show the presence of an excess of oxygen, the amount remaining is determined by absorption with potassium pyrogallate.

The calculation is given on page 43. For very accurate work a second analysis should be made, making successive explosions, using the percentages of methane and hydrogen just found as a basis upon which to calculate the quantity of oxygen to be added each time. The explosive mixture should be so proportioned that the ratio of combustible gas (i.e., CH_4, H and O) is to the gases which do not burn (i.e., N and the excess of CH_4 and H) as 100 is to about 50 (from 26 to 64);* otherwise the heat developed is so great as to produce oxides of nitrogen, which, being absorbed

* Bunsen, Gasometrische Methoden, 2d ed., p. 73 (1877).

in the potassium hydrate, would affect the determination of both the methane and the hydrogen. The oxygen should preferably be pure, although commercial oxygen, the purity of which is known, can be used; the oxygen content of the latter should be tested from time to time, especially with different samples.

(*b*) *Hempel's Method.**—From 12 to 15 cc. of the gas are measured off into the burette (e.g., 13.2 cc.) and the residue is passed into the cuprous chloride pipette for safe keeping. That in the burette is now passed into the small explosion pipette; a volume of air more than sufficient to burn the gas, usually about 85 cc., is accurately measured and also passed into the explosion pipette, and in so doing water from the burette is allowed to partially fill the capillary of the pipette and act as a seal. The rubber connectors upon the capillaries of the burette and pipette are carefully wired on, both pinch-cocks shut, and the stop-cock closed. The pipette is cautiously shaken, the screen interposed, the explosion wires connected with the induction coil, a spark passed between them, and the stop-cock immediately opened. The gas in the pipette, containing carbon dioxide, oxygen, and nitrogen, is transferred to the burette, accurately measured, by reading immediately, to prevent the absorption of carbon dioxide, and carbon dioxide and oxygen determined in the usual way.

Calculation.—(*a*) *Hinman's Method.*—56.2 cc. of gas remained after the absorptions; 77.4 cc. of oxygen were introduced, giving a total volume of 133.6 cc.

---

* Hempel, Gas Analytische Methoden, 3d ed., p. 245 (1901).

Residue after explosion........... 46.9 cc.
Residue after CO_2 absorption...... 28.2 "
Carbon dioxide formed............. 18.7 "
Contraction.......133.6 — 46.9 = 86.7 "
Residue after O absorption......... 25.6 "
Oxygen in excess, 28.2 — 25.6 =   2.6 "

The explosion of marsh-gas or methane is represented by the equation *

$$\boxed{CH_4} + \boxed{O_2}\ \boxed{O_2} = \boxed{CO_2} + \boxed{H_2O} + \boxed{H_2O}.$$

From this it is evident that the volume of carbon dioxide is equal to the volume of methane present; therefore in the above example, in the 56.2 cc. of gas burned there were 18.7 cc. methane.

The total contraction is due (1) to the disappearance of oxygen in combining with the hydrogen of the methane, and (2) to the union of the free hydrogen with oxygen. The volume of the methane having been found, (1) can be ascertained from the equation above, equals twice the volume of the methane; hence

$$86.7 - (2 \times 18.7) = 49.3 \text{ cc.},$$

contraction which is due to the combustion of hydrogen. This takes place according to the following reaction: *

$$\boxed{H_2} + \boxed{H_2} + \boxed{O_2} = \boxed{H_2O} + \boxed{H_2O}.$$

---

* H_2O being as steam at 100° C. At ordinary temperatures this is condensed, giving rise to "total contraction."

Hydrogen then requires for its combustion half its volume of oxygen, hence this 49.3 cc. represents a volume of hydrogen with $\frac{1}{2}$ its volume of oxygen, or $\frac{3}{2}$ volumes; hence the volume of hydrogen is 32.9 cc.

(*b*) *Hempel's Method.*—Of the 82 cc. of gas remaining after the absorptions, 13.2 cc. were used for the explosion; 86.4 cc. air introduced giving a total volume of 99.6 cc.

> Residue after explosion............ 78.0 cc.
> Residue after CO_2 absorption...... 73.2 "
>
> Carbon dioxide formed............ 4.8 "
> Contraction........99.6 − 78.0 = 21.6 "
> Residue after O absorption........ 70.2 "
> Oxygen in excess..73.2 − 70.2 =   3.0 "

The carbon dioxide being equal to the methane present, in the 13.2 cc. of gas burned, there were 4.8 cc. of methane. The volume of methane is found by the proportion 13.2 : 82 :: 4.8 : 4, whence $x =$ 29.8 cc.

The hydrogen is calculated similarly.

The following method of calculation may be substituted for that on page 43: Let m = methane, h = hydrogen, c = total contraction, and O = oxygen actually used; then

$$2m + \frac{h}{2} = O$$

and

$$2m + \frac{3h}{2} = c,$$

whence

$$m = \frac{3O - c}{4}$$

and

$$h = c - O.$$

The explosion can also be made after the absorption of oxygen and thus the troublesome absorption of carbonic oxide avoided. The calculation is then, if $C =$ carbonic oxide, $K = CO_2$ formed:

$$c = \frac{C}{2} + 2m + \frac{3h}{2}, \quad . \quad . \quad . \quad . \quad (1)$$

$$K = C + m, \quad . \quad . \quad . \quad . \quad . \quad (2)$$

$$V = C + m + h; \quad . \quad . \quad . \quad (3)$$

whence

$$h = V - K,$$

$$C = \frac{K}{3} + V - \frac{2c}{3},$$

$$m = \frac{2K}{3} - V + \frac{2c}{3}.$$

Another method for the estimation of hydrogen is by absorption with palladium sponge;[*] it, however, must be carefully prepared, and it is the author's experience that one cannot be sure of its efficacy when it is desired to make use of it. A still better absorbent of hydrogen[†] is a 1 per cent solution of palladous

---

[*] Hempel, Berichte deutsch. ch. Gesell., **12**, 636 and 1006 (1879).
[†] Campbell and Hart, Am. Chem. J., **18**, 294 (1896).

chloride at 50° C.; when fresh this will absorb 20–50 cc. of hydrogen in ninety minutes. A proportionately longer time is required if more hydrogen be present or the solution nearly saturated. The methane could then be determined by explosion or by mixing with air and passing to and fro over a white-hot platinum spiral in a tubulated pipette called the grisoumeter * (grisou = methane).

Nitrogen.—There being no direct and convenient method for its estimation with this apparatus, the percentage is obtained by finding the difference between the sum of all the percentages of the gases determined and 100 per cent.

New † determines nitrogen in illuminating-gas directly after the method of Dumas in organic substances; 150 cc. of gas are used, the hydrocarbons partially absorbed by fuming sulphuric acid and the remainder burned in a combustion tube with copper oxide; the carbon dioxide is absorbed and the residual nitrogen collected and measured.

Accuracy and Time Required.—For the absorptions the apparatus is accurate to 0.1 cc.; for explosions by Hinman's method ‡ the methane can be determined within 0.2 per cent, the hydrogen within 0.3 per cent; by Hempel's method within 1 per cent for the methane and 7.5 per cent for the hydrogen. The time required for the analysis of illuminating-gas is from three to three and one-half hours; for air, from fifteen to twenty minutes.

---

* Winkler, Fres. Zeit., **28**, 269 and 288.
† J. Soc. Chem. Ind., **11**, 415 (1892).
‡ Gill and Hunt, *loc cit.*

Notes.—The object in filling the capillaries of the explosion pipettes with water or mercury before the explosion is to prevent the bursting of the rubber connectors on them. With mercury this is effected by introducing it through the T joint in the connector. After testing for oxygen with the pyrogallate a small quantity of dilute acetic acid is sucked into the burette to neutralize any alkali which by any chance may have been sucked over into it. The acid is rinsed out with water and this forced out by mercury before the burette is used again.

The water in the burette should be saturated with the gas which is to be analyzed—as illuminating-gas —before beginning an analysis. The reagents in the pipettes should also be saturated with the gases for which they are not the reagent. For example, the fuming sulphuric acid should be saturated with oxygen, carbon monoxide, methane, hydrogen, and nitrogen; this is effected by making a blank analysis using illuminating-gas.

The method of analysis of the residue after the absorptions have been made by explosion is open to two objections: 1st, the danger of burning nitrogen by the violence of the explosion; and 2d, the danger of breakage of the apparatus and possible injury to the operator. These may be obviated by employing the apparatus of Dennis and Hopkins,* which is practically a grisoumeter with mercury as the confining liquid; or that of Jager, † who burns the gases with oxygen in a

---

* J. Am. Chem. Soc., **21**, 398 (1899).

† J. f. Gasbeleuchtung, **41**, 764. Abstr. J. Soc. Chem. Ind., **17**, 1190 (1898).

hard-glass tube filled with copper oxide. By heating to 250° C. nothing but hydrogen is burned; higher heating of the residue burns the methane. Or the mixture of oxygen and combustible gases, bearing in mind the ratio mentioned at the bottom of page 43; can be passed to and fro through Drehschmidt's * capillary heated to bright redness. This consists of a platinum tube 20 cm. long, 2 mm. thick, 1.7 mm. bore, filled with three platinum or palladium wires. The ends of the tube are soldered to capillary brass tubes and arranged so that these can be water cooled. It is inserted between the burette and a simple pipette, mercury being the confining liquid in both cases. The air contained in the tube can be determined as in the case of the tube containing iodic anhydride, p. 42.

To the method of explosion by the mixture of an aliquot part of the residue with air, method (*b*), there is the objection that the carbon dioxide formed is measured over water in a moist burette, giving abundant opportunities for its absorption, and that the errors in anylysis are multiplied by about six, in the example by $\frac{820}{132}$.

---

* Ber. d. deut. chem. Gesell. **21,** 3242 (1888).

CHAPTER VI.

REAGENTS AND ARRANGEMENT OF THE LABORATORY.

THE reagents used in gas-analysis are comparatively few and easily prepared.

Hydrochloric Acid, Sp. gr. 1.10.—Dilute "muriatic acid" with an equal volume of water. In addition to its use for preparing cuprous chloride, it finds employment in neutralizing the caustic solutions which are unavoidably more or less spilled during their use.

Fuming Sulphuric Acid.—Saturate "Nordhausen oil of vitriol" with sulphuric anhydride. Ordinary sulphuric acid may be used instead of the Nordhausen; in this case about an equal weight of sulphuric anhydride will be necessary. *Absorption capacity*, 1 cc. absorbs 8 cc. of ethene (ethylene).

Acid Cuprous Chloride.—The directions given in the various text-books being troublesome to execute, the following method, which is simpler, has been found to give equally good results. Cover the bottom of a two-liter bottle with a layer of copper oxide or " scale " ⅜ in. deep, place in the bottle a number of pieces of rather stout copper wire reaching *from top to bottom,* sufficient to make a bundle an inch in diameter, and fill the bottle with common hydrochloric

acid of 1.10 sp. gr.   The bottle is occasionally shaken,
and when the solution is colorless, or nearly so, it is
poured into the half-liter reagent bottles, containing
copper wire, ready for use.   The space left in the
stock bottle should be immediately filled with hydrochloric acid (1.10 sp. gr.).

By thus adding acid or copper wire and copper
oxide when either is exhausted, a constant supply of
this reagent may be kept on hand.

The absorption capacity of the reagent per cc. is,
according to Winkler, 15 cc. CO; according to
Hempel 4 cc.   The author's experience with Orsat's
apparatus gave 1 cc.

Care should be taken that the copper wire does not
become entirely dissolved and that it extend from the
top to the bottom of the bottle; furthermore the
stopper should be kept thoroughly greased the more
effectually to keep out the air, which turns the solution
brown and weakens it.

Ammoniacal Cuprous Chloride. — The acid cuprous chloride is treated with ammonia until a faint
odor of ammonia is perceptible; copper wire should
be kept in it similarly to the acid solution.   This
alkaline solution has the advantage that it can be
used when traces of hydrochloric acid vapors might
be harmful to the subsequent determinations, as, for
example, in the determination of hydrogen by absorption with palladium.   It has the further advantage
of not soiling mercury as does the acid reagent.

Absorption capacity, 1 cc. absorbs 1 cc. CO.

Cuprous chloride is at best a poor reagent for the
absorption of carbonic oxide; to obtain the greatest

accuracy where the reagent has been much used, the gas should be passed into a fresh pipette for final absorption, and the operation continued until two consecutive readings agree exactly. The compound formed by the absorption—possibly Cu_2COCl_2—is very unstable, as carbonic oxide may be freed from the solution by boiling or placing it in vacuo ; even if it be shaken up with air, the gas is given off, as shown by the increase in volume and subsequent diminution when shaken with fresh cuprous chloride.

Hydrogen.—A simple and effective hydrogen generator can be made by joining two six-inch calcium chloride jars by their tubulatures. Pure zinc is filled in as far as the constriction in one, and the mouth closed with a rubber stopper carrying a capillary tube and a pinch-cock. The other jar is filled with sulphuric acid 1 : 5 which has been boiled and cooled out of access of air. The mouth of this jar is closed with a rubber stopper carrying one of the rubber bags used on the simple pipettes.

Mercury.—The mercury used in gas analysis should be of sufficient purity as not to "drag a tail" when poured out from a clean vessel. It may perhaps be most conveniently cleaned by the method of J. M. Crafts, which consists in drawing a moderate stream of air through the mercury contained in a tube about 3 feet long and $1\frac{1}{4}$ inches internal diameter. The tube is supported in a mercury-tight V-shaped trough, of size sufficient to contain the metal if the tube breaks, one end being about 3 inches higher than the other. Forty-eight hours' passage of air is sufficient to purify any ordinary amalgam. The mercury may very well be kept in a large separatory funnel under a layer of strong sulphuric acid.

Palladous Chloride.—5 grams palladium wire are dissolved in a mixture of 30 cc. hydrochloric and 2 cc. nitric acid, this evaporated just to dryness on a water-bath, redissolved in 5 cc. hydrochloric acid and 25 cc. water, and warmed until solution is complete. It is diluted to 750 cc. and contains about one per cent of palladous chloride. It will absorb about two thirds of its volume of hydrogen.

Phosphorus.—Use the ordinary white phosphorus cast in sticks of a size suitable to pass through the opening of the tubulated pipette.

Potassium Hydrate.—(*a*) For carbon dioxide determination, 500 grams of the commercial hydrate is dissolved in 1 liter of water.

Absorption capacity, 1 cc. absorbs 40 cc. CO_2.

(*b*) For the preparation of potassium pyrogallate for special work, 120 grams of the commercial hydrate is dissolved in 100 cc. of water.

Potassium Pyrogallate.—Except for use with the Orsat or Hempel apparatus, this solution should be prepared only when wanted. The most convenient method is to weigh out 5 grams of the solid acid upon a paper, pour it into a funnel inserted in the reagent bottle, and pour upon it 100 cc. of potassium hydrate (*a*) or (*b*). The acid dissolves at once, and the solution is ready for use.

If the percentage of oxygen in the mixture does not exceed 28, solution (*a*) may be used;* if this amount be exceeded, (*b*) must be employed. Otherwise carbonic oxide may be given off even to the extent of 6 per cent.

* Clowes, Jour. Soc. Chem. Industry, **15**, 170.

Attention is called to the fact that the use of potassium hydrate purified by alcohol has given rise to erroneous results.

Absorption capacity, 1 cc. absorbs 2 cc. O.

Sodium Hydrate.—Dissolve the commercial hydrate in three times its weight of water. This may be employed in all cases where solution (*a*) of potassium hydrate is used. The chief advantage in its use is its cheapness, it costing but one tenth as much as potassium hydrate, a point to be considered where large classes are instructed. Sodium pyrogallate is, however, a trifle slower in action than the corresponding potassium salt.

ARRANGEMENT OF THE LABORATORY.

The room selected for a laboratory for gas-analysis should be well lighted, preferably from the north and east. To prevent changes in temperature it should be provided with double windows, and the method of heating should be that which will give as equable a temperature as possible. In the author's laboratory, instead of the usual tables, shelves are used, 18 inches wide and $1\frac{1}{4}$ inches thick, best of slate or soapstone, firmly fastened to the walls, 30 inches from the floor; the Orsat apparatus, when not in use, may be suspended from these. The reagents are contained in half-liter bottles fitted with rubber stoppers, placed upon a central table convenient to all. Here are found scales, funnels and graduates for use in making up reagents. Distilled water is piped around to each place by $\frac{1}{8}$-inch tin pipe and $\frac{3}{16}$-inch rubber tubing from a $\frac{1}{4}$-inch "main," being supplied at the tem-

perature of the room from bottles placed about six
feet above the laboratory shelves. A supply of a
gallon per day per student should be provided.

At the right of each place is fixed a sand-glass of
cylindrical rather than conical form, graduated to
minutes for the draining of the burettes. The "egg-
timers" found in kitchen-furnishing stores serve the
purpose admirably.

"Unknown gases" for analysis are best contained
in a Muencke double aspirator, Fig. 14, where they

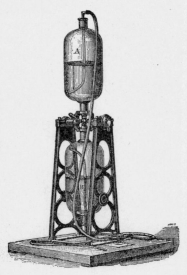

Fig. 14.—Muencke's Aspirator.

can be thoroughly mixed before distribution and con-
veyed by a pipe to the central table.

Finally, the laboratory should contain a stone-ware
sink provided with an efficient trap of the same

material, to prevent mercury from being carried into and corroding the lead waste-pipes.

Drawers should be provided with compartments for various sizes of rubber connectors, pinchcocks, glass tubing, stoppers and fittings, and tools. When working with the Orsat apparatus alone, three feet of shelf space may be allowed to each student; when using this with another, as, for example, the Bunte, another foot should be added.

The course which the writer has been in the habit of giving to the Mechanical and Electrical Engineers embraces two exercises in the laboratory of two hours each, supplemented with four hours of lectures. The students in the laboratory make an analysis of air and an " unknown " furnace-gas, take and analyze an actual sample of chimney-gas, and make the calculation of heat lost and air used. In the lectures, the subject of gas-analysis and its other applications, and of fuels, their origin, description, preparation, analysis, and determination of heating value, are described.

CHAPTER VII.

FUELS—SOLID, LIQUID, AND GASEOUS: THEIR DERIVATION AND COMPOSITION.

The substances employed as fuels are:

a. SOLID FUELS.—Wood, peat, brown, bituminous and anthracite coal, charcoal, coke, and oftentimes various waste products, as sawdust, bagasse, straw, and spent tan.

b. LIQUID FUELS.—Crude petroleum and various tarry residues.

c. GASEOUS FUELS.—Natural gas, producer, blast-furnace, water, and illuminating gas.

The essential constituents in all these are carbon and hydrogen; the accessory, oxygen, nitrogen, and ash; and the deleterious, water, sulphur, and phosphorus.

a. SOLID FUELS.

Wood is composed of three substances—*cellulose*, or woody fibre $(C_6H_{10}O_5)_n$; *the components of the sap*, the chief of which is lignine, a resinous substance of identical formula with cellulose; and *water*. The formation of cellulose from carbon dioxide and water may be represented by the equation

$$6CO_2 + 5H_2O = C_6H_{10}O_5 + 6O_2.$$

The amount of water which wood contains determines its value as a fuel. This varies from 29 per cent in ash to 50 per cent in poplar; it varies also with the season at which the wood is cut, being least when the sap is in the roots—in December and January. This difference may amount to 10 per cent in the same kind of wood.

The harder varieties of wood make the best fuel, a cord of seasoned hardwood being about equal to a ton of coal. Yellow pine, however, has but half this value; the usual allowance in a boiler-test is 0.4 the value of an equal weight of coal.

The ash of wood is mainly potassium carbonate, with traces of other commonly occurring substances, as lime, magnesia, iron, silica, and phosphoric acid.

The *percentage composition of wood* may be considered as approximately,

Water.	Carbon.	Hydrogen.	Oxygen.	Ash.	Sp. Gr.
20	39	4.4	35.6	1	0.5.*

When burned it yields about 4000 C. per kilo, and requires 6 times its weight of air or 4.6 cu. m. (74.1 cu. ft. per pound) for its combustion.

Peat finds considerable application in Europe, and is coming into use in this country in the form of briquettes. To this end it is reduced to a dry powder and compressed into small cylindrical blocks; it is claimed to be as efficient as coal at half the price. It is also proposed to gasify peat after the manner of coal. Peat is produced by the slow decay under water of certain swamp plants, more especially the mosses (Sphagnaceæ), evolving methane (CH_4) (marsh-gas) and carbon dioxide (CO_2).

It contains considerable moisture, from 20 to 50 per cent, and 10 per cent even when "thoroughly

---

* Mills & Rowan, Fuels, p. 11.

dry." Thirty per cent of its available heat is employed in evaporating this moisture. The high content of ash, from 3 to 30 per cent, averaging 15 per cent, also diminishes its value as a fuel.

The ash of peat differs from that of wood in containing little or no potassium carbonate.

The *percentage composition of peat* may be considered as approximately,

	Water.	Carbon.	Hydrogen.	Oxygen.	Nitrogen.	Ash.	Sp. Gr.
German....	16.4	41.0	4.3	23.8	2.6	11.9	1.05
American...	20.8	40.8	4.4	26.6		7.7	—

Such peat is about equivalent to wood in its heating effect, one pound evaporating from 4.5 to 5 pounds of water.

Coal.—Geologists tell us that coal was probably produced by the decay under fresh water of plants belonging principally to the Conifer, Fern and Palm families; these flourished during the Carboniferous Age to an extent which they never approached before or since. Representatives of the last family, which it is thought produced most of the coal, have been found 2 to 4 feet in diameter and 80 feet in height.

By their decay, carbon dioxide " choke-damp," marsh-gas " fire-damp," and water were evolved. The change might be represented by the equation

$$6C_6H_{10}O_5 = 7CO_2 + 3CH_4 + 14H_2O + C_{26}H_{20}O_2.$$

Cellulose.   Bituminous Coal.

Some idea of the density of the vegetation and the time required may be obtained from the fact that it has been calculated that 100 tons of vegetable matter —the amount produced per acre per century—if compressed to the specific gravity of coal and spread over an acre would give a layer less than 0.6 of an inch thick. Now four fifths of this is lost in the evolution of the gaseous products, giving as a result an accumulation of *one eighth of an inch per century*, or one foot in 10,000 years.*

Brown Coal or Lignite may be regarded as forming the link between wood and coal; geologically speaking it is of later date than the true coal. Most of the coal west of the Rocky Mountains is of this variety.

As its name denotes, it generally is of brown color —although the western coal is black—and has a conchoidal fracture. It contains a large quantity of water when first mined, as much as 60 per cent, and when " air-dry " from 15 to 20 per cent. The per cent of ash is also high, from 1 to 20 per cent.

The average moisture and ash in American lignites are 12.75 and 6.1 respectively.

The *percentage composition of brown coal* may be considered as approximately,

	Water.	Carbon.	Hydrogen.	Oxygen & Nitrogen.	Ash.	Sp. Gr.
German	18.0	50.9	4.6	16.3	10.2	1.3

Bituminous Coal.—This is the variety from which all the following coals are supposed to have been formed, by a process of natural distillation combined with pressure. According to the completeness of this process we have specimens which contain widely differing quantities of volatile matter. This forms the true basis for the distinguishing of the varieties of coal. In ordinary bituminous coal this volatile matter amounts to 30 or 40 per cent. Three varieties of bituminous coal are ordinarily distinguished, as follows:

---

* In case the student desires to follow in a more extended manner the geology of coal, reference may be had to Le Conte's " Elements of Geology," pp. 345–414, 3d ed.

Dry or non-caking —those which burn freely with but little smoke and—as the name denotes—do not cake together when burned. The coals from Wyoming are an example of this class.

Caking—those which produce some smoke and cake or sinter together in the furnace. An example of these is the New River and Connellsville coal.

Fat or Long-flaming—those producing much flame and smoke and do or do not cake in burning; volatile matter 50 per cent or more. Some of the Nova Scotia coals belong to this class.

Bituminous coal varies much in its composition—is black or brownish black, soft, friable, lustrous, and of specific gravity of 1.25 to 1.5.

Moisture varies from 0.25 to 8 per cent, averaging about 5.

The *percentage composition of bituminous coal* may be considered as approximately,*

Water.	Carbon.	Hydrogen.	Oxygen.	Nitrogen.	Ash.	Sulphur.
0.9	77.1	5.2	6.7	1.6	7.6	1.0

Water.	Volatile Matter.	Fixed Carbon.	Ash.
0.9	27.4	64.1	7.6

Semi-Bituminous or Semi-Anthracite Coal is upon the border-line between the preceding and the following variety; it is harder or softer than bituminous, contains less volatile matter (15 to 20 per cent), and burns with a shorter flame. An example of this is the Pocahontas coal.

The *percentage composition of semi-bituminous and semi-anthracite coal* may be considered to be approximately,*

Water.	Carbon.	Hydrogen.	Oxygen.	Nitrogen.	Ash.	Sulphur.
0.5	83.0	4.7	4.2	1.3	5.5	0.8

Water.	Volatile Matter.	Fixed Carbon.	Ash.
0.5	16.7	77.3	5.5

* H. J. Williams.

Anthracite Coal is the hardest, most lustrous, and densest of all the varieties of coal, having a specific gravity of 1.3 to 1.75; it contains the most carbon and least hydrogen and volatile matter (5 to 10 per cent). It has a vitreous fracture and kindles with difficulty, burning with a feeble flame, giving little or no smoke and, with sufficient draft, an intense fire. The Lehigh coal is an excellent example of this class.

The *percentage composition of anthracite coal* may be considered as approximately,*

Water.	Carbon.	Hydrogen.	Oxygen.	Nitrogen.	Ash.	Sulphur.
2.0	83.9	2.7	2.8	0.8	7.2	0.6

Water.	Volatile Matter.	Fixed Carbon.	Ash.
2.0	4.3	86.5	7.2

The ash of coal varies from 1 to 20 per cent and is mainly clay—silicate of alumina—with lime, magnesia, and iron. When coal is burned it yields from 6100 to 8000 C. and requires about 12 times its weight of air, 9.76 cu. m. per kilo or 156.7 feet per pound. For the greatest economy Scheurer-Kestner † found that this should be increased from 50 to 100 per cent.

Charcoal is prepared by the distillation or smouldering of wood, either in retorts, where the valuable by-products are saved, or in heaps. It should be jet-black, of bright lustre and conchoidal fracture.

When wood is charred in heaps only about 20 per cent of its weight in charcoal is obtained—48 bushels per cord, or about half the percentage of carbon. When retorts or kilns are employed, the yield is increased to 30 per cent, and 40 per cent of pyroligneous

---

* H. J. Williams.
† Jour. Soc. Chem. Industry, **7**, 616.

acid of 10 per cent strength, with 4 per cent of tar, are obtained.

The *percentage composition of wood-charcoal* may be considered as approximately,

Carbon.	Ash.	Sp. Gr.
97.0	3.0	0.2

Coke is prepared by the distillation of bituminous coal in ovens; these are of two types, those in which the distillation-products are allowed to escape—the " beehive " ovens—and those in which they are carefully saved, as the Otto-Hoffman, Semet-Solvay, Simon-Carvés', and others.

The "beehive" ovens yield from 50 to 65 per cent of the weight of the coal—about $2\frac{1}{2}$ tons. The Otto-Hoffman ovens are long narrow thin-walled retorts 33 by 6 by 1.5 feet,[*] regeneratively heated by side and bottom flues; the charge is about 6 tons of coal, and the following percentage yields of by-products are obtained: coke 70–75, gas 16 (10 M. cu. ft.), tar 3.3–5.6, ammonia 0.3–1.4.[†] The Semet-Solvay ovens differ from the above in that they are not regeneratively heated and their walls are thicker, serving to store up the heat; the yield of coke is somewhat higher—about 80 per cent.[‡] The by-products obtained alone increase the value of the output about one and one half times. Good coke should possess a cellular structure, a metallic ring, contain practically no impurities, and be capable of bearing a heavy burden in the furnace.

---

[*] Irwin, Eng. Mag., Oct. 1901, abstr. J. Am. Chem. Soc., **24**, 40.
[†] H. O. Hofman, Tech. Quar., **11**, 212 (1898).
[‡] Pennock, J. Am. Chem. Soc., **21**, 678 (1899).

The analysis of Connellsville coke with the coal from which it is prepared is given below.

	Water.	Volatile Matter.	Carbon.	Sulphur.	Ash.
Coal	1.26	30.1	59.62	0.78	8.23
Coke	0.03	1.29	89.15	0.084	9.52

Otto-Hoffman coke:

		Fixed Carbon.	
3.7	1.3	86.1	8.9

Heating value...................... 7100 C.

The Minor Solid Fuels.

Sawdust and **Spent Tan-bark** find occasional use, their value depending upon the quantity of moisture they contain. With 57 per cent of moisture 1 pound of tan-bark gave an evaporation of 4 pounds of water.

Wheat Straw finds application as fuel in agricultural districts, 3½ pounds being equal to 1 pound of coal. Upon sugar-plantations the crushed cane or **Bagasse,** partially dried, is extensively used as a fuel. With 16 per cent of moisture an evaporation of 2 pounds of water per pound of fuel has been obtained.

Briquets, "Patent Fuel." *—In Europe coal dust is cemented together with some tarry binding material and baked or compressed into blocks usually about $6 \times 2 \times 1$ inches, which form a favorite fuel for domestic purposes. In some cases they take the form and size of a large goose egg, and are called eggettes: these are being made, among other places, at Scranton, Pa., and withstand well the shocks incident to shipment.

---

* Condition of the Coal Briquetting Industry in the United States, E. W. Parker, Bull. No. 316 U. S. Geol. Survey, Contributions to Economic Geology, 1906. Part II, Coal Lignite and Peat, pp. 460–485.

Storage of Coal aud Spontaneous Combustion.—
While authorities differ as to the way and manner in
which coal should be stored, as regards height of pile,
number, size, and arrangement of ventilating channels,
they are practically agreed that it should always be
covered. Six months' exposure to the weather may with
European coals cause a loss of from 10 to 40 per cent in
heating value, while with Illinois coals it varies from
2 to 10 per cent.* The North German Lloyd Steamship
Company stores its coal in a covered bin provided with
ventilators, and restricts the height of the pile to 8 feet.
A large gas company in a western city also uses a covered
bin, with ventilators 8 inches square every 20 feet; the
height of the pile may be from 10 to 15 feet. An electric
company in the same city † has arranged to store 14,000
tons of coal under water in 12 pits, a steam-shovel being
used to dig out the coal. Ventilating flues serve the
additional purposes of enabling the temperature of the
pile to be ascertained before ignition takes place, and as
a means of introduction of either steam or carbonic acid
to extinguish any fire which may occur. All the supports
of the bin in contact with the coal should be of brick,
concrete or iron, and if of hollow iron, filled with cement.

The spontaneous combustion of coal is due primarily
to the rapid absorption of oxygen by the finely divided
coal, and to the oxidation of iron pyrites, "coal brasses,"
occurring in the coal. The conditions favorable to
the process are:

---

* Parr and Hamilton, Univ. of Ill., Bulletin 4, No. 33, August,
1907.

† Eng. and Min. Jour., September 15, 1906.

First. A supply af air sufficient to furnish oxygen, but of insufficient volume to carry off the heat generated.

Second. Finely divided coal, presenting a large surface for the absorption of oxygen.

Third. A considerable percentage of volatile matter in the coal.

Fourth. A high external temperature.

A method of extinguishing a fire in a coal pile not provided with ventilators consists in removing and spreading out the coal and flooding the burning part with water. Another method consists in driving a number of iron or steel pipes provided with "driven well points" at the place where combustion is taking place, and forcing water or steam through these upon the fire.

b. Liquid Fuels.

These consist of petroleum and its products, and various tarry residues from processes of distillation, as from petroleum, coking-ovens, wood and shale. Liquid fuel possesses the advantage that it contains no ash, is easily manipulated, the fire is of very equable temperature, very hot, and practically free from smoke.

Regarding the origin of petroleum, many theories have been proposed. That of Engler,* that it was formed by the distillation under pressure of animal fats and oils, the nitrogenous portions of the animals previously escaping as amines, seems most probable; it has yielded the best results of any hypothesis when tested upon an industrial scale.

---

* Jour. Soc. Chem. Industry, 14, 648.

Crude Petroleum varies greatly in color according to the locality; it is usually yellowish, greenish, or reddish brown, of benzine-like odor, and sp. gr. of 0.78 to 0.80. It " flashes " at the ordinary temperature: hence great care should be employed in its use and storage. Its *percentage composition* is shown below.

Carbon.	Hydrogen.
84.0–85.0	16.0–15.0

It is more than twice as efficient as the best anthracite coal. In practice 14 to 16 pounds of water per pound of petroleum have been evaporated, and an efficiency of 19,000 B. T. U. was obtained as against 8500 B. T. U. for anthracite. In general $3\frac{1}{2}$ to 4 barrels of oil are equivalent to a ton of good soft coal.*

c. GASEOUS FUELS.

Natural Gas is usually obtained when boring for petroleum and consists mainly of methane and hydrogen, although the percentage varies with the locality. The Findlay, Ohio,† gas is of the following composition:

CH_4	H	N	O	C_2H_4	CO_2	CO	H_2S	Sp. Gr.
92.6	2.3	3.5	0.3	0.3	0.3	0.5	0.2	0.57

Blast-furnace, Producer, or Generator Gas is the waste gas issuing from the top of a blast-furnace or obtained by partially burning coal by a current of air (pro-

---

* W. B. Phillips, Texas Petroleum (1900), p. 84.
† Orton, Geology of Ohio, vol VI. p. 137.

duced by steam) in a special furnace—a gas-producer or generator. It is mainly carbonic oxide and nitrogen.

	CO	N	CO_2	H	CH_4	O	B.T.U. per Foot.
Blast-furnace gas.....	34 3	63.7	0.6	1.4	—	—	—
Gas from bitumin. coal	24.5	46.8	3.7	17.8	6.8	0.4	223
" " " "	25.0	41.4	4.0	19.4	9.6	0.6	—
" " anthrac. "	27.0	57.3	2.5	12.0	1.2	—	—
" " " "	17.2	53.1	8.6	18.2	2.4	0.4	140
" " " "	26.0	47.0	8.0	18.5	0.5	—	145 *

One ton of coal yields from 160† to 170 thousand cubic feet of gas of 156 to 138 B.T.U. heating power, or 81 to 86 or even 90 per cent of the value of the coal.

Water-gas.—If, instead of passing simply air over hot coal, water-vapor, or rather steam, be employed, it is decomposed, giving carbonic oxide and hydrogen, according to the equation $H_2O + C = CO + H_2$, and the resulting mixture is called water-gas. The *percentage composition*, which varies according to the apparatus and fuel employed, is about as follows:

	CO	H	CH_4	CO_2	N	O	Illts.	Sp. Gr.
From coke....	45.8	45.7	2.0	4.0	2.0	0.5	—	0.57
From bit. coal	34.0	41.9	7.5	5.4	9.2	1.1	0.9	—

Fischer ‡ states that 1 ton of coke gives about 36 thousand cubic feet of gas, equivalent to 42 per cent of the value of the coal. From 1 ton of bituminous coal about 51 thousand cubic feet of gas of 360 B.T.U. heating power are obtained, or an efficiency of nearly 62 per cent.§

---

* Suction gas producer.
† Humphrey, Jour. Soc. Chem. Industry, **20**, 107 (1901); *ibid.*, **16**, 522 (1897).
‡ Taschenbuch für Feuerungs-Techniker, p. 27.
§ Slocum, J. Soc. Chem. Industry, **16**, 420 (1897).

Coal or Illuminating Gas was formerly produced by the distillation of bituminous coal; it is at present largely made by the enriching of water-gas. "Gas-oil," a crude naphtha, is blown into the water-gas generator and changed to a permanent gas by the heat.   It is of the following composition:

	H	CH_4	CO	C_2H_4	CO_2	N	O	Sp. Gr.
Coal gas............	47.0	40.5	6.0	4.0	0.5	1.5	0.5	0.4
Enriched water-gas	27.9	25.9	25.3	15.0	2.9	3.0	0.0	0.6

One ton of coal gives about 10 thousand cubic feet of gas, or about 20 per cent of the heating value of the coal.

Heating Value of these Gases.

The following table, mainly from Slocum,* gives an idea of the comparative value of the gases:

Name of Gas.	B. T. U. per Cu. Ft.†	Yield.	Air Required *for* Combustion per Cu. Ft.
Oil..................	1350	77 cu. ft. per gal.	....
Natural.............	980	....	9.80
		Thousand Ft. per Ton.	
Enriched water......	700	40 ‡	....
Coke-oven..........	686	5	....
Coal................	600–625	10	5.65
Blue water..........	500	5	....
Heating (coke-oven).	367	5	....
Bit. coal water......	342	51	2.97
Mond producer......	156	160	1.25
Siemens producer....	137	170	....

* Slocum, J. Soc. Chem. Industry, **16**, 420 (1897).
† Determined with the Junkers calorimeter.
‡ 168–200 gallons of "gas-oil" are also required.

REFERENCES.—*Report of U. S. "Liquid Fuel" Board*, Dept. of Navy, Bureau of Steam Engineering, Washington, 1904. pp. 450.

Report on the Operations of the Coal-Testing Plant of the U. S. Geological Survey at St. Louis, 1904. Professional Paper, No. 48, Parts I, II, and II. 1906.

Preliminary Report on the Operations of the Coal Testing Plant of the U. S. Geol. Survey at St. Louis, 1904.

Bull. No. 261, 1905.

Bull. No. 261 for 1905.

Bull. No. 290, 1906.

A Study of Four Hundred Steaming Tests, made at Fuel Testing Plant at St. Louis in 1904, 1905, 1906, by L. P. Breckenridge. Bull. No. 325, U. S. Geol. Survey, 1907.

The Burning of Coal without Smoke. D. T. Randall. Bull. No. 334, U. S. Geol. Survey, 1908.

Barr, *"Boilers and Furnaces."*

Hodgetts, *" Liquid Fuels."*

CHAPTER VIII.

METHODS OF ANALYSIS AND DETERMINATION ON THE HEATING VALUE OF FUEL.

SAMPLING.

A FEW representative lumps or shovelfuls are taken from each barrow or from various points in the pile in boiler tests. Shovelfuls of coal should be taken at regular intervals and put into a *tight covered* barrel or some air-tight receptacle, and the latter should be placed where it is protected from the heat of the furnace.* In sampling two conditions must be observed: First, the original sample should be of considerable size and thoroughly representative; and, second, the quartering down to an amount which can be put into a sealed "lightning" jar should be carried out as quickly as possible after the sample is taken. Careful samplings and careful treatment of samples are necessary to obtain reliable results, especially in the determination of moisture. The lumps are coarsely broken, and the whole spread out in a low circular heap. Diameters are drawn at right angles in it and opposite quarters taken, and treated similarly to the whole sample. The operation is continued until a sample of a few pounds is obtained. This is roughly crushed and samples taken at different points for the moisture determination; it is then further quartered down until a

---

* Report of Committee on Coal Analysis, J. Am. Chem. Soc., **21**, 1116 *et seq.* (1899).

sample of 100 grams which passes a 60-mesh sieve is obtained.

The methods employed in the analysis of fuels are largely a matter of convention, various methods giving varied results; for example, it is well-nigh impossible to obtain accurately the percentage of moisture in coal, as when heated sufficiently hot to expel the water some of the hydrocarbons are volatilized.

Moisture.—Dry one gram of coal in an open crucible at 104°–107° C. for one hour. Cool in a desiccator and weigh covered. Where accuracy is required, determinations must also be made on the coarsely ground sample; this latter result is to be regarded as the true amount and corrections applied to all determinations where the powdered sample is used.* †

Volatile Combustible Matter and Coke.*§—Place one gram of fresh, undried powdered coal in a platinum crucible having a tightly fitting cover. Heat over the full flame of a Bunsen burner for seven minutes by the watch. The crucible should be supported on a platinum triangle with the bottom six to eight centimeters above the top of the burner. The flame should be fully twenty centimeters high when burning free, and the determination should be made in a place free from drafts. The upper surface of the cover should burn clear, but the under surface should remain covered with carbon. To find "Volatile Combustible Matter" subtract the per cent of moisture from the loss found here. The residue in the crucible minus the ash represents the *Coke or Fixed Carbon.*

---

* Report of Committee on Coal Analysis, *loc. cit.*

† See also an article by Hale, Proc. Am. Soc. Mech. Eng. 1896.

§ Sommermeier, J. A. C. S. **28**, 1002 (1906).

Certain non-coking coals suffer mechanical loss from the rapid heating.

Carbon and Hydrogen.—These are determined by burning the coal in a stream of air and finally in oxygen, the products of combustion, carbon dioxide and water, being absorbed in potassium hydrate and calcium chloride.

Apparatus Required.—Combustion-furnace similar to that shown in Fig. 15. Combustion-tube filled.

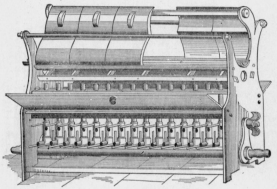

FIG. 15.—COMBUSTION-FURNACE.

Potash-bulbs with straight chloride of calcium tube filled. Chloride of calcium tube filled. Oxygen-holder, drying and purifying apparatus. Porcelain boat, desiccator, tongs, $\frac{1}{8}$-inch rubber tubing. Analytical balance.

The combustion-tube is of hard glass, $\frac{1}{2}$ inch in internal diameter and 36 inches long, closed with perforated rubber stoppers. One end—called the front end—is filled with a layer of copper oxide 12 inches long, held in place by plugs of asbestos coming within 4 inches of the stopper. In coals rich in sulphur the oxide is partially replaced by a layer of chromate of lead 2 inches long. The position of the boat containing the coal is immediately behind this copper oxide; behind the boat is placed an oxidized copper gauze roll, 6 inches long. Before making the combustion, the tube and contents should be heated to a dull red heat in a stream of oxygen freed from moisture and carbon dioxide by the purifying apparatus, to burn any dust and dry the contents; it is then ready for use.

The potash-bulbs are an aggregation of five bulbs, the three lowest filled with potassium hydrate of 1.27 sp. gr., the other two serving as safety-bulbs, preventing the liquid from being carried over into the connectors. They should be connected further with a chloride of calcium tube to absorb any moisture carried away by the dry gas. When not in use they should be closed with connectors carrying glass plugs. Before weighing they should stand at least fifteen minutes in the balance-room to attain its temperature; the weight should be to milligrams and without the connectors.

The chloride of calcium tube is of U form, provided with bulbs for the condensation of the water; the granular calcium chloride is kept in place by cotton plugs, and the stopper neatly sealed in with sealing-wax. As calcium chloride may contain oxide which would absorb the carbon dioxide formed, a current of dry carbon dioxide should be passed through the tube and thoroughly swept out by dry air before use.

The chloride of calcium tube like the potash-bulbs should be placed in the balance-room fifteen minutes before weighing and, if the balance-case be dry, may be weighed without the connectors. It should be weighed to milligrams.

The oxygen-holder may be like the Muencke aspirator, Fig. 14. The oxygen should be purified by passing through potassium hydrate and over calcium chloride.

Operation.—The front stopper of the combustion-tube is slipped carefully upon the stem of the chloride of calcium tube and this connected to the potash-bulbs; 0.2 to 0.3 gram of the coal is carefully weighed into the porcelain boat (to 0.1 mg.), the roll removed, and the boat inserted behind the layer of copper oxide, and the roll and stopper replaced. The tube is now ready to be heated.

The front of the copper oxide is first heated, the heat being gradually extended back; at this time the rear end of the copper roll is heated and a slow current of purified air passed through. This method of gradual heating of the tube is followed until the layer of copper oxide and the rear portion of the roll are at a dull red heat. Heat is now cautiously applied to the coal and the current of air slackened. The volatile matter in the coal distils off, is carried into the layer of copper oxide and burned; the carbon dioxide formed can be seen to be absorbed by the potassium hydrate. When this absorption almost ceases, oxygen is turned on and the coal heated until it glows. The stream of oxygen should be so regulated as to produce but two bubbles of carbon dioxide in the bulbs per second. If the evolution be faster, the gas is not absorbed. When the coal has ceased glowing, oxygen is allowed to pass through the apparatus until a spark held at the exit of the last chloride of calcium tube (on the bulbs) re-inflames; the oxygen is allowed to run for fifteen minutes longer. The current of oxygen is now replaced by purified air, and the heat moderated by turning down the burners and opening the fire-clay tiles; the air is allowed to run through for twenty minutes to thoroughly sweep out all traces of carbon dioxide and moisture. The bulbs and U tube are disconnected, stopped up, allowed to stand in the balance-room, and weighed as before. The increase in weight in the bulbs represents the carbon dioxide formed; this multiplied by the factor 0.2727 gives the carbon. Similarly the increase in the U tube, minus the water due to the moisture in the coal, represents the water formed, one ninth of which is hydrogen.

Notes.—At no time in the combustion should any water appear near the copper roll, as it is an indication that the products of combustion have gone backward into the purifying apparatus and hence are lost. Such analyses should be repeated. Should moisture appear in the front end, it may be *gently* heated to expel it. Both ends of the tube should be frequently touched with the hand during the combustion, and should be no hotter than may be comfortably borne, as the stoppers give off absorbable gases when highly heated. Care should be taken not to heat the tube too hot, fusing the copper oxide into and spoiling it. One tube should serve for a dozen determinations. It should not be placed upon the iron trough of the furnace, but upon asbestos-paper in the trough, to prevent fusion to the latter.

As will be seen, the execution of a combustion is not easy, and should only be intrusted to an experienced chemist. The results obtained are usually 0.1 per cent too low for carbon and a similar amount too high for hydrogen.

Ash.—This is determined by weighing the residue left in the boat after combustion, or by completely burning one gram of the coal contained in a platinum dish; often a stream of oxygen is used.

Nitrogen is determined by Kjeldahl's method, which consists in digesting the coal with strong sulphuric acid, aided by potassium permanganate, until nearly colorless. The nitrogenous bodies are changed to ammonia, which forms ammonium sulphate and may be determined by rendering alkaline and distilling the solution.

Sulphur is determined by Eschka's method, consisting in heating for an hour one gram of the coal mixed with one gram of magnesium oxide and 0.5 grm. sodium carbonate in a platinum dish without stirring, using an alcohol-lamp, as gas contains sulphur. It is allowed to cool and rubbed up with one gram of ammonium nitrate and heated for 5 to 10 minutes longer. The resulting mass is dissolved in 200 cc. of water evaporated to 150 cc., acidified with hydrochloric acid, filtered, and sulphuric acid determined in the filtrate in the usual way with barium chloride.

Oxygen is determined by difference, there being no direct method known.

ANALYSIS OF LIQUID FUELS.

Carbon and Hydrogen.—This determination is made as in the case of the solid fuels, the liquid being contained in a small bulb sealed for weighing to prevent volatilization. The stem is scratched and broken off and the bulb inserted in the combustion tube in place of the boat. Extra care in heating has to be observed to prevent the liquid from passing through unburnt. For thick or tarry oils having a small quantity of volatile matter, the boat may be used as with solid fuels.

Sulphur.—For oils containing more than 0.01 per cent sulphur the well known method of Carius may be employed. This consists in sealing up the oil contained in a small weighing tube, in a tube with fuming nitric acid and barium chloride and heating in a furnace for several hours. All sulphur is converted into sulphuric acid, which combines with the barium chloride forming barium sulphate, which is filtered off and weighed in the usual way. Another method consists in burning the oil in a small lamp and collecting the products of combustion. The lamp is a miniature "oil lamp" made from a 3-inch test-tube (weighing tube) by drawing a piece of lamp wicking through a small glass tube contained in the stopper. This lamp is suspended by a wire from the balance and weighed accurately.

It is lighted and hung under a funnel arranged so that the products of combustion are drawn by an air-pump through a series of two washing bottles containing saturated bromine water. The lamp is weighed after about a gram of oil has been burned, the bromine boiled out, the solution evaporated to about 150 cc., acidified with hydrochloric acid and the sulphuric acid formed determined in the usual way with barium chloride.

Nitrogen is determined exactly as in the case of solid fuels.

Water can be shown qualitatively by the eosine test * by rubbing with a little eosine on a glass plate. If water be present the oil will take on a pink color. For its quantitative determination a weighed amount of gently ignited plaster of Paris is added to the oil and allowed to stand 24–36 hours. Gasoline is now added to the oil and the whole brought upon a dried weighed filter and the plaster washed until all oil is removed; the filter and contents are dried at a gentle heat not exceeding 100° C. to a constant weight. The increase in weight represents the quantity of water in the oil.

Flash and Fire Test.—Determined by heating the oil in the covered New York tester, according to Gill, "Short Handbook of Oil Analysis," Fourth Edition, pp. 12–14.

The analysis of gaseous fuels has already been described in Chapter V.

---

* Holley and Ladd, Mixed Paints, Color Pigments and Varnishes, p. 36.

DETERMINATION OF CALORIFIC POWER OF SOLID AND LIQUID FUEL.

a. Direct Methods.

Many forms of apparatus have been proposed for this purpose; few, however, with the exception of those employing Berthelot's principle—of burning the substance under a high pressure of oxygen—have yielded satisfactory results. The apparatus of William Thomson,[*] and also that of Barrus, in which the coal is burnt in a bell-jar of oxygen, while usually yielding results within 3 per cent of the calculated value, yet they may vary as much as 8 per cent from that value.[†] Unless a crucible lined with magnesia be used, or the sample mixed with bituminous coal, it is inapplicable to certain semi-bituminous and anthracite coals, as the ash formed over the surface prevents the combustion of the coal beneath it.

Fischer's calorimeter [‡] is similar in principle, but is claimed to give very good results.[§]

Lewis Thompson's calorimeter, in which the coal is burnt in a bell-jar by the aid of oxygen furnished by the decomposition of potassium chlorate or nitrate, is open to several objections, the chief of which are:

1. The evolution of heat due to the decomposition of the

---

[*] Thomson, Jour. Soc. Chemical Industry, 5, 581.
[†] Ibid., 8, 525.
[‡] Zeit. f. angewandte Chemie, 12, 351.
[§] Bunte, Jour. f. Gasbeleuchtung und Wasserversorgung, 34, 21, 41.

oxidizing substance used. 2. Loss of heat due to moisture carried off by the gases in bubbling through the water. The results which it gives must be increased by 15 per cent.*

Hempel's apparatus † makes use of the Berthelot principle: the coal must be compressed into a cylinder for combustion—a process to which every coal is not adapted — only applicable to certain varieties of bituminous and brown coal. The mixture with the coal of any cementing or inflammable substance to form these cylinders carries with it the necessity of accurately determining its calorific power beforehand.

The best apparatus for the purpose is probably that of Mahler,‡§ modified by Williams, Norton, and Emerson, the modifications consisting in replacing the enamel lining by a nickel one or by electroplating the inside with gold and in improved methods of making the apparatus tight.

The Mahler apparatus, Fig. 17, consists of a mild-steel cylinder *B*, with walls half an inch thick, narrowed at the top for connection by a screw-joint with the cover carrying the vessel *C* to contain the coal. This cylinder or bomb is placed inside the calorimeter *D*, and this inside a jacket *A*. At the right is shown a portion of the oxygen-cylinder and the gauge.

For the following directions for its use the author is indebted to the kindness of Professor Silas W. Holman of the Institute of Technology.

---

* Scheurer-Kestner Jour. Soc. Chemical Industry, **7**, 869.
† Hempel, "Gasanalytische Methoden," p. 347.
‡ Mahler, Jour. Soc. Chemical Industry, **11**, 840.
§ Mayer, Stevens Indicator, (1895) **134**.

Preparation of Bomb.—Remove the ring upon which it sits in the calorimeter.

Wash out the bomb. It need not be dry. Leave cover off.

See that the lead-ring washer P, Fig. 16, is in good

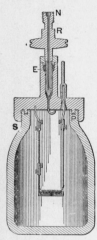

condition. Unless its upper surface is fairly smooth the cover cannot be tightly closed. Repeated screwing on of the cover raises a burr of lead. When this becomes noticeable it must be removed by cutting with a knife-blade. If there is difficulty in making the cover tight, it is most likely to be due to this cause.

Grease the screw S upon the outside of the bomb slightly with tallow or a heavy oil, but be sure that none of the grease gets beyond the lead washer.

FIG. 16.—MAHLER'S BOMB.

Secure the bomb very firmly in the heavy clamp on the table.

Place the top on a ring or in a clamp of a lamp-stand and in an upright position.

Put in position the platinum tray C and the rod E, Fig. 17.

Twist on the loop of ignition-wire (fine platinum or iron). This must make good electrical contact with both E and the pan or its supporting rod. Failing this the current will not flow to fuse the wire. Failure to ignite is almost always traceable to this cause.

Pour into the tray a known weight of the substance

to be burned. If this be coal, slightly over one gram should be used. It is usually best inserted from a small test-tube weighed before and after, with due precautions against loss.

The ignition-wire should dip well into the coal.

The fineness required in the combustible depends

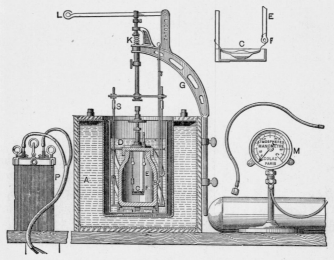

FIG. 17.--MAHLER'S APPARATUS COMPLETE.

upon its nature. Anthracite coal should be in a very fine powder, at least 100 mesh. Trial will show whether any unburned grains remain, indicating that the combustible is too coarse.

The standard which carries the pressure-gauge should be screwed to the table near the bomb-clamp, and the oxygen cylinder must be placed near by so that the three may be easily connected by the flexible copper tube.

The top carrying the charge is then cautiously (to avoid loss of charge by jarring or draft) transferred to the bomb and screwed carefully home. The lifting is best done by hooking the fingers beneath the milled head at the top of the valve-screw R. The top must be set up hard by the wrench which takes the large nut cut on the cover. In setting this up it is desirable to use no more force than is necessary to secure a gas-tight bearing of the tongue of the cover against the lead washer P. Just the force required can only be learned by experience, but it is always considerable. A slight leak is unimportant, but it is not difficult to secure a tight seal if the lead washer be kept in good condition.

To fill with oxygen proceed as follows:

Screw down the valve-screw R *gently* to close the valve. Connect the copper tube to the oxygen-tank gauge, and to the bomb at N. See that there are leather washers at the joints. Turn the connecting nuts firmly but not violently home. The connections to the oxygen-tank and gauge are usually left undisturbed, and only that at N has to be made each time.

It is now necessary to test for leakage in the connections. To do this, as R is closed, it is only necessary to open the oxygen-tank *cautiously* by means of its wrench until the gauge indicates 5 or 10 atmospheres and then close it. As the tank when freshly charged has a pressure of 120 atmospheres, and the gauge reads only to 35 atmospheres, *care must be used in all manipulations not to overstrain the gauge*, also avoid suddenly releasing the pressure on the gauge. When this pressure is on, any leak in the connections will be indicated by a drop in the gauge reading. If a leak exists, it must be removed or rendered extremely slow before proceeding further. It is most likely to be found in the joints, which must be tightened one by one until the leak stops.

Now to fill the bomb it is next necessary to open R. This could be done by merely turning back the milled head, or the nut just above it. But as this would put a twist into the copper connecting-tube (which many times repeated would break it), the better way is, holding one wrench in each hand, to loosen the connecting nut above N *by a half-turn*, holding R by the wrench and nut, then to turn the nut open a half-turn or until it is again tight in. This leaves the connections tight and R open into the bomb. The oxygen is then turned slowly on, and the bomb gradually fills. If a gram of coal is to be burned, a pressure of 25 atmospheres gives the proper amount of gas in the bomb. Note that the valve R and the inlet-tube have small borings. Thus the inflow of gas will be slow and the pressure in the connecting-tube will be higher than in the bomb. If, therefore, the tank be closed quickly, the gauge-reading will fall somewhat until these pressures equalize, and will then remain stationary unless there is a leak. The tank-cock must always be kept well under control to avoid overcharging either gauge or bomb.

When the bomb is full, close first the tank-cock. Then, to close R, put the wrenches on the nuts and, holding one from turning, set the other down until R is tight, *but not too tight*. Avoid straining R, which closes tight *very easily*. By this method the copper

tube is not twisted. There is of course a slight leak
of gas from the bomb after N leaves the nut and
before R is closed, but the time required for the half-
turn is so short and the outflow so slow that the loss
is insignificant. There is *no need to hurry* in this
operation. Be deliberate and careful of the apparatus.
A valve like R is a nice piece of workmanship, and to
endure much usage it must be treated with care.

The bomb is now ready to be unclamped and set
into the ring preparatory to transfer to the calorimeter.
It can be left standing indefinitely, but must be
handled with caution (best by lifting with fingers
beneath R, to avoid spilling the charge).

Preparation of Calorimeter.—The outer jacket of
the calorimeter should be filled with water at about
the room temperature or a few degrees higher. If
left standing from day to day it will usually be nearly
enough right. It is well to stir it (blow air through
it) somewhat before beginning work, if it has stood for
some time.

Be sure that the inner surface of this jacket, i.e.,
the one which is next the calorimeter, is *thoroughly
dry*, and do not let any water spill into it—or remove
it if it does so.

Thoroughly dry the outer surface of the calorimeter
and *keep it so*. Moisture depositing on or evaporating
from the surface of the calorimeter is sure to cause an
irregular error which may spoil otherwise good work.

Put the calorimeter in place. Transfer the bomb to
it, and adjust the stirrer so that it works properly.

Pour in the proper amount of water, about 2.25

liters, at a suitable temperature, best by using marked flasks carefully calibrated beforehand.

Insert the thermometer.

See that the electrical attachments are ready for instantaneous use. The whole is then ready for the combustion.

Combustion Observations.—With apparatus all in place run the stirrer briskly and continuously until the completion of the work. Allow about five minutes for everything to come to a normal condition. Then take temperature readings to at least 0.01° at each quarter minute for at least five minutes. Record the times (h. m. s.) and corresponding thermometer-readings, thus:

Time.	Temp.	Remarks.
$2^h \ 15^m \ 0^s$	$15°.24$	After 5^m stirring
15	.24	
30	.25	
45	.25	
16 0	.25	
15	.26	
30	.26	
. . . .		
25 0	—	Coal ignited
15	15.6	
30	.9	
45	16.2	
26 0	.5	
etc.	etc.	

Exactly at the beginning of a noted minute close the electric circuit through the fuse-wire. If the arrangements are right, this will cause the coal to ignite at once and the combustion is almost instantaneous. Owing to the time required to transmit the heat through the bomb to the water, the temperature, however, will continue to rise for two or three minutes. Keep up the steady stirring and the quarter-minute temperature-readings for at least ten minutes after ignition, recording as above. One or two observations may be unavoidably lost before and after ignition, but this does not materially affect the results. The readings during the rapid rise are also less close.

As soon as the rise begins to slow down, however, the hundredths of a degree must again be secured. This makes a series of observations of 15 to 20 minutes' duration. The use of the readings to obtain the cooling correction and the corrected rise of temperature of the calorimeter is given under the heading " Cooling Correction " farther on.

This completes the observations unless it is desired to test the character of the products of combustion. The bomb should now be opened and rinsed, as the nitric acid formed by the oxidation of the nitrogen in the coal and air attacks the metallic lining unless it be of gold. Also the top is more easily unscrewed at first than later. Leave the top off.

Before unscrewing the top of the bomb *be sure to open the valve R* to relieve the presure.

Heat Capacity of Bomb and Calorimeter.—The heat capacity of the bomb may be found:

1. From the weights and assumed specific heats of the parts.

2. By raising the bomb to an observed high temperature and immersing in water, i.e., by the usual " method of mixtures."

3. By burning in it a substance of known heat of

combustion, such as pure naphthaline, and calculating back to find the heat capacity of the bomb.

The first method is not reliable. Errors of several per cent may enter in the assumed specific heats.

The second method is very difficult of exact performance, owing to the size and form of the bomb.

The third method is by far the most reliable, but of course depends on the correctness of the assumed heat of combustion of the substance used. That of naphthaline has been so well determined by Berthelot and others,* and the substance is so easily and cheaply obtained in a pure state, that dependence can be placed on the results. This method has the great advantage that it involves the use of the apparatus in precisely the same way as in subsequent determination, so that any systematic errors of method tend to cancel one another. It also determines at the same time the heat capacity of the calorimeter and stirrer just as used.

The capacity of the calorimeter and stirrer may best be determined in connection with that of the bomb by the third method just described. Otherwise it may be found by the first method, or by a method similar to the second, viz., by pouring into the calorimeter when partly full water of a known temperature different from that of the water in the calorimeter, noting all temperatures and weights. This last method, however, is very unsatisfactory in practice owing to the small heat capacity of the calorimeter and to the losses of heat in pouring the water, etc.

---

* 1 gram of naphthaline evolves 9692 C. This is the average of 150 determinations by four different obervers.

A general expression for computing the heat of combustion from the bomb observations is as follows: Let n represent the number of grams of combustible, H the heat of combustion sought, W the weight of water in the calorimeter, and k the heat capacity, or water equivalent, of bomb, calorimeter, stirrer, thermometer, etc.; t_1 and t_2 represent the initial and final temperatures of the water. Then

$$nH = W(t_2 - t_1) + k(t_2 - t_1),$$

whence

$$H = \frac{1}{n}(W + k)(t_2 - t_1).$$

This expression is exact if t_2 is corrected for loss by cooling as described in the methods for " Cooling Correction," p. 85.

The value of k may be determined by either of the following methods; a simplification may, however, be introduced which will save much labor if an accuracy of not more than about one per cent is sought, provided that k is found by burning naphthaline or other known substance. Use enough of this substance to cause about the same rise, $t_2 - t_1$, (within 1°) as will be caused by one gram of coal. Omit the cooling correction entirely, using for t_2 the maximum temperature attained. Then compute k; this value will be erroneous by a small amount owing to the neglect of the correction. Now in subsequent measurements on coal also neglect the cooling correction, using for t_2 the maximum observed temperature as before, thus leaving an error in t_2. Since the rise $t_2 - t_1$ in both cases will be nearly the same, the error in k will almost exactly affect that in t_2 in the coal-test, and the result-ing value of H will be nearly free from this error. This method of course implies that W is nearly constant and that t_1 is systematically arranged to be either about at the air-temperature or a definite amount below it, as described under " Cooling Correction," so that the cooling loss is about the same. The time-interval from t_1 to t_2 must for the same reason be nearly constant in all cases.

Cooling Correction.—In all careful calorimetric work, one of the most troublesome sources of error is the loss or gain of heat by the calorimeter from its surroundings. This loss or gain is due to radiation, to air-convection currents, and to evaporation or condensation. Unavoidable irregularities in the conditions and the smallness of the quantities to be measured render the amount of the correction variable and its determination uncertain. Many methods of making the correction have been proposed. One of the best of these is the first of the two given below, but the second, although a little more troublesome in the execution of the work, appears to be more trustworthy in its results. The second method is to be used.

First Method.—This is described in the Physical Laboratory Notes, I,* under " Specific Heat of Solids." In this method the water at the outset should be at such a temperature that it is gaining very slowly. For an open calorimeter this is about $1°$ or $2°$ below the air-temperature, but varies with circumstances. Water which has been long standing in the room is generally about right.

Second Method.—For the discussion and details

---

* Obtainable from A. D. Machlachlan, Bookseller, Boston.

reference may be had to an article by Professor Holman
in Proc. American Academy of Arts and Sciences, 1895,
p. 245; also in The Technology Quarterly, **8**, 344.

Parr's Calorimeter.*—This '' has the advantage of
operating without an oxygen gas supply; its manipulation is simple and the extraction of the heat rapid,
owing to the compact mass in which the heat is generated. It is especially adapted to soft coal, and
while designed for technical purposes, its factor of
error is well within 0.5 per cent.'' ''It depends for its
action upon the liberation of oxygen from a compound
which shall in turn absorb the products of combustion,
conditions admirably met in
sodium peroxide; this ob-
viates the necessity of pro-
viding for an outlet for those
gases and also any loss aris-
ing from the heat they
might carry off.''

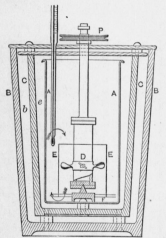

''*A*, Fig. 18, is the calorimeter of about two liters
capacity, insulated by two
outer vessels of indurated
fiber, *B* and *C*, so placed as
to provide further insulation
by the air-spaces *b* and *c*.

FIG. 18.—PARR'S CALORIMETER. The cover is double, to correspond, with an air-space between, the two parts
being connected for convenience in handling. The
cartridge *D* has a capacity of about 25 cc.; it rests
on a pivot below, extends through the covers, and has
a small removable pulley at the end. Turbine wings

fastened to spring clips are placed on the cartridge, and a short cylinder E, open at both ends, is provided for directing the current set up by rotation of the vanes attached to the cartridge. The stem G of the cartridge is so arranged as to permit the passage of a short piece of No. 12 copper wire to ignite the charge; it is provided with a valve D at the lower end to prevent the escape of the enclosed air."

Manipulation.—One gram of coal ground to pass a 100-mesh sieve, dried at 105°, is put into the cartridge, 16–18 grams of sodium peroxide added, the top screwed on, and the whole shaken to thoroughly mix the contents. The peroxide should be fine enough to pass a 25-mesh sieve. The cartridge is tapped to settle the charge to the bottom, placed in the calorimeter, two liters of water poured in, and rotated 50 to 100 revolutions per minute. The water should be 3 to 4 degrees lower than the room temperature. When the temperature has become constant, the thermometer is read, a hot wire dropped down G, igniting the charge, which burns completely. The extraction of the heat is effected in about five minutes; the reading of the maximum temperature is taken and the calculations made as follows: The rise of temperature is corrected, first, for that produced by the hot wire; this amounts to 0.006° C. per ¼ inch of No. 12 copper wire: second, for the heat produced by the combination of the sodium peroxide with the carbon dioxide and water formed by the combustion; this amounts to 27 per cent of the total indicated heat. If $C =$ the heat of combustion of the coal, C' the calories indicated, t the rise of temperature, and w the water employed, then

$$C' = (t - 0.006°) \times w, \quad C = C' - \frac{C' \times 27}{100},$$

$$C = (t - 0.006°) \times w \times 0.73.$$

Notes.—Instead of using a gram of coal some prefer to use half this quantity, mixing it thoroughly and rapidly with the peroxide in a watch-glass with a spatula and transferring it to the combustion-chamber.

Ignition by an electrically heated platinum wire is to be preferred to that by dropping a hot copper wire into the mixture. Accidents have been caused from the failure of the valve to work.

The combustion-chamber should be perfectly dry within and without.

Ashes and coke are difficult of ignition: this can be effected by adding a second charge of peroxide with half a gram of good coal, the combustion factor of which has been determined, and thoroughly mixing the charges and repeating the ignition.

For stirring, the smallest size electric or water motor furnishes sufficient power.

In the formulæ $w =$ the water employed $+$ the water equivalent of the calorimeter.

Berthier's Method. — Another method of direct determination was proposed by Berthier in 1835.[*] It uses as a measure of the heating value the amount of lead which a fuel would reduce from the oxide; in other words, it is proportional to the amount of oxygen absorbed.

The method is as follows[†]: Mix one gram of the

---

[*] Dingler's Polytechnisches Journal, **58**, 391.
[†] Noyes, McTaggart & Craver, J. Am. Chem. Soc., **17**, 847 (1798).

finely powdered dry coal with 60 grams of oxide of lead (litharge) and 10 grams of ground glass. This mixing can be done with a palette-knife on a sheet of glazed paper; the mixture is transferred to a fire-clay crucible (Battersea C size), covered with salt, the crucible covered and heated to redness in a hot gas-furnace—or the hottest part of the boiler-furnace—for 15–20 minutes. After cooling, the crucible is broken and the lead button carefully cleaned and weighed. Multiply the weight of the lead button obtained by 268.3 calories (or 483 B. T. U.) and divide the product by the weight of coal taken. The result is the number of calories per gram or B. T. U. per pound. One gram of lead is theoretically equivalent to 234 calories (C); owing to the hydrogen present this factor gives results about two per cent too low. The results obtained by the author using " horn-pan " scales in one case by this method were within 2.8 per cent of those yielded by the bomb calorimeter, which are as close as those obtained by any calorimeter save Parr s. The method would seem worthy of more attention than it has received.

b. *Determination of Heating Value by Calculation.*

The method of determination of the heating value first described, though exact, has the disadvantages that the apparatus is costly and the compressed oxygen is not easily obtained. To obviate these, it has been sought to obtain the heating value by calculation from the chemical analysis, the heating value of the constituents being known. This has the disadvantage that we have no absolute knowl-

edge—nay, not even an approximate idea—as to how
the carbon, hydrogen, water, and sulphur exist in the
coal, so that any formula must of necessity be quite
removed from the truth. Dulong was the first to
propose the method by calculation, and his formula * is

$$H = \frac{8000c + 34500(h - \frac{1}{8}o)}{100},$$

c, h, and o representing the percentages of carbon,
hydrogen, and oxygen in the coal.

Many modifications of this, considering the water
formed, the heat of vaporization of carbon, or the
volatile hydrocarbons, have been proposed.

Bunte † finds that the following formula ‡ gives results
varying from + 2.8 to − 3.7 per cent:

$$\frac{8080c + 28800\left(h - \frac{o}{8}\right) + 2500s - 600w}{100}.$$

s and w represent the percentages of sulphur and
water respectively. It is, however, inapplicable to
anthracite coal. It would scarcely seem that the
sulphur would be worth considering unless high, one
per cent affecting the result but 0.3 per cent.

Mahler employs the formula *

$$\frac{8140c + 34500h - 3000(o + n)}{100},$$

o and n representing oxygen and nitrogen, and states
that it gives results within 3 per cent.

The results obtained by these formulæ for anthracite
coal are as a rule considerably too low.

---

* H burned to liquid water.   ‡ H burned to aqueous vapor.
† Jour für Gasbeleuchtung, **34**, 21-26 and 41-47.

Goutal * has proposed the formula

$$P = \frac{8150c + AM}{100}$$

as being more readily applicable than the preceding.

P represents the calorific power; *c*, the percentage of fixed carbon (coke − ash) ; *M*, the percentage of volatile matter (100 − [coke + ash + water]) ; *A* is a coefficient which varies with the amount of volatile matter *M*, viz. :

$M =$	2 to 15	$A =$	13000
	15 to 30		10000
	30 to 35		9500
	35 to 40		9000

The results upon a series of American coals varied less than 2 per cent from those obtained by the calorimeter.

REFERENCES.—An admirable discussion of the errors incident to the use of the Mahler calorimeter and others of that type will be found in a paper by G. A. Fries, Bulletin No. 94 (1907), U. S. Dept. of Agriculture, Bureau of Animal Industry, "Investigations in the Use of the Bomb Calorimeter."

---

* Revue de chimie ind., **7** (1896), 65; abs. in The Analyst, **24**, 107.

CALORIFIC POWER OF GASEOUS FUEL.

a. Direct Determination.

Perhaps the best apparatus for the determination of the heating value of gases is the Junkers calorimeter, Figs. 19 and 20. The following description is taken

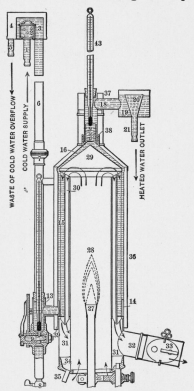

FIG. 19.—JUNKERS GAS-CALORIMETER (SECTION).

from an article by Kühne in the Journal of the Society of Chemical Industry, vol. 14, p. 631. As will be

seen from Fig. 19, this consists of a combustion-chamber, 28, surrounded by a water-jacket, 15 and 16,
this being traversed by a great many tubes. To
prevent loss by radiation this water-jacket is sur-

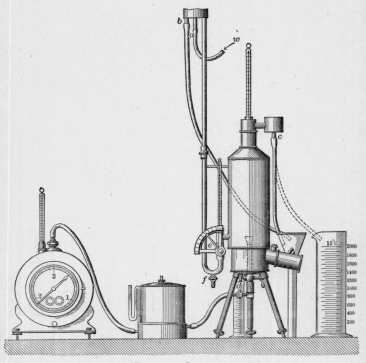

FIG. 20.—JUNKERS GAS-CALORIMETER.

rounded by a closed annular air-space, 13, in which
the air cannot circulate. The whole apparatus is
constructed of copper as thin as is compatible with
strength. The water enters the jacket at 1, passes
down through 3, 6, and 7, and leaves it at 21, while

the hot combustion-gases enter at 30 and pass down, leaving at 31. There is therefore not only a very large surface of thin copper between the gases and the water, but the two move in opposite directions, during which process all the heat generated by the flame is transferred to the water, and the waste gases leave the apparatus approximately at atmospheric temperature. The gas to be burned is first passed through a meter, Fig. 20, and then, to insure constant pressure, through a pressure-regulator. The source of heat in relation to the unit of heat is thus rendered stationary; and in order to make the absorbing quantity of heat also stationary, two overflows are provided at the calorimeter, making the head of water and overflow constant. The temperatures of the water entering and leaving the apparatus can be read by 12 and 43; as shown before, the quantities of heat and water passed through the apparatus are constant. As soon as the flame is lighted, 43 will rise to a certain point and will remain nearly constant.

Manipulation.—The calorimeter is placed as shown in Fig. 2c, so that one operator can simultaneously observe the two thermometers of the entering and escaping water, the index of the gas-meter, and the measuring-glasses.

No draft of air must be permitted to strike the exhaust of the spent gas.

The water-supply tube *w* is connected with the nipple *a* in the centre of the upper container; the other nipple, *b*, is provided with a waste-tube to carry away the overflow, which latter must be kept running while the readings are taken.

The nipple *c* through which the heated water leaves the calorimeter is connected by a rubber tube with the large graduate. *d* empties the condensed water into the small graduate.

The thermometers being held in position by rubber stoppers and the water turned on by *e* until it discharges at *c*, no water must issue from *d* or from 39, Fig. 19, as this would indicate a leak in the calorimeter.

The cock *e* is now set to allow about two liters of water to pass in a minute and a half, and the gas issuing from the burner ignited. Sufficient time is allowed until the temperature of the inlet-water becomes constant and the outlet approximately so; the temperature of the inlet-water is noted, the reading of the gas-meter taken, and at this same time the outlet-tube changed from the funnel to the graduate. Ten successive readings of the outflowing water are taken while the graduate (2-liter) is being filled and the gas shut off.

A better procedure is to allow the water to run into tared 8-liter bottles, three being used for a test, and weighing the water. The thermometer in the outlet can then be read every half-minute.

EXAMPLE.—Temp. of incoming water, 17.2°
    " " outgoing " 43.8°
                  ―――――
    Increase,   26.6°
    Gas burned, 0.35 cu. ft.

$$\text{Heat} = \frac{\text{liters water} \times \text{increase of temp.}}{\text{cu. ft. gas}} = \frac{2 \times 26.6}{0.35}$$

$$= 152\ 3\text{C}.$$

From burning one cubic foot of gas 27.25 cc. of water were condensed. This gives off on an average 0.6 C. per cc.

$$27.25 \times 0.6 = 16.3 \text{ C.};$$
$$152.3 - 16 \text{ } 1 = 136 \text{ C. per cubic foot;}$$
$$136 \times 3.96823 = 540 \text{ B. T. U.}$$

The calorific power obtained without subtracting the heat given off by the condensation of the water represents the *total* heating value of the gas. This is the heat given off when the gas is used for heating water or in any operation where the products of combustion pass off below 100° C. The *net heating value* represents the conditions in which by far the greater quantity of gas is consumed, for cooking, heating and gas engines, and is the one which should be reported. It should, however, be corrected, as shown on page 99, to the legal cubic foot, that is, measured at 30 inches barometric pressure, and 60° F. saturated with moisture.

The apparatus has been tested for three months in the German Physical Technical Institute with hydrogen, with but a deviation of 0.3 per cent from Thomson's value. This value may vary nearly that amount from the real value owing to the method which he employed.

b. By Calculation.

Oftentimes it may be impracticable to determine the heating value of gases directly; in such cases recourse must be had to the calculation of its calorific power from volumetric analysis of the gas.

To this end multiply the percentage of each constituent by its number as given in Table IV, and the sum of the products will represent the British Thermal Units evolved by the combustion of one cubic foot of the gas.* It is assumed that the temperature of the gas burned and the air for combustion is 60° F., and that of the escaping gases is 328° F., that corresponding to the temperature of steam at 100 pounds absolute pressure.

As has been already stated, column 3 in Table IV is based upon the assumption that the gas, and air for its combustion, enter at 60° F., and the products of combustion leave at 328° F.; in column 4 it is assumed that the entering temperature of both gas and air is 32° F., and the combustion-gases are cooled to 32° F. In case these conditions are varied, the amount of heat which the gas and air bring in must be determined; this is found in the usual way by multiplying the proportionate parts of 1 cubic foot, as shown by the analysis, by the specific heat of the gas, and this by the rise in temperature (difference between observed temperature and 32° F.). The quantity of air necessary for combustion is found by multiplying the percentage composition of the gas by the number of cubic feet necessary for the combustion of each constituent.

An example will serve to make this clear. The analysis of Boston gas is as follows:*

CO_2	"Illuminants."	O	CO	CH_4	H	N
2.9	15.0	0.0	25.3	25.9	27.9	3.0

Or in one cubic foot there are

---

* H. L. Payne, Jour. Analytical and Applied Chem., **7**, 230.

† Jenkins, Annual Report Inspector of Gas Meters and Illuminating Gas, 1896, p. 11.

.029 CO_2 .... .259 CH_4

.150 "illuminants" .... .279 H

.253 CO .... .030 N

Let us assume that the gases, instead of passing out at a temperature of 328° F., leave at the same temperature as that of the chimney-gases, p. 29, 250° C. or 482° F.

The calculation of the heat carried away is similar to that there given.

0.15 cu. ft. of "illuminants" produces, Table III, 0.3 cu. ft. CO_2 and 0.3 cu. ft. steam;

0.253 cu. ft. of carbonic oxide produces .253 cu. ft. CO_2;

0.259 cu. ft. methane produces 0.259 cu. ft. CO_2 and .518 cu ft. steam;

0.279 cu. ft. hydrogen produces .279 cu. ft. steam.

From the combustion of the gas there results .812 cu. ft. CO_2, 1.097 cu. ft. steam, and 5.90 × 79.08 or 4.665 cu. ft. N.

The quantity of heat they carry off is as follows:

	Vol.		Vol. Sp. Ht.		Rise.		B.T.U.
CO_2	.812	×	.027	×	450	=	9.9
N	4.66	×	.019	×	450	=	39.9
Excess of air ..	1.2	×	.019	×	450	=	10.2
Steam.........	1.097	×	.0502	×	1229	=	67.7

Total heat lost............... = 127.7

The loss due to the steam is found by multiplying the weight of steam found by the "Total Heat of Steam," as found from Steam Tables.* The tables, however, do not extend beyond 428° F.; it can be calculated by the formula

$$\text{Total heat} = \lambda = 1091.7 + 0.305(t - 32).$$

One cubic foot of hydrogen when burned yields .0502 lbs. of water.

The heat generated by the combustion of the gas is found by multiplying its volume by its calorific power, Table IV.

"Illuminants" .... $0.15 \times 2000.0 = 300.0$ B.T.U.
CO............... $0.253 \times 341.2 = 86.3$
CH_4............... $0.259 \times 1065.4 = 276.0$
H............... $0.279 \times 345.4 = 96.3$

Heat generated by the gas.......... 758.6 B.T.U.

Total heat lost (p. 98)....... 127.7

630.9 B.T.U.

This figure, 630.9 B.T.U., represents the heating power of one cubic foot of the gas measured at 62° F., and is consequently too small; its heating value at 32° F. is represented by

$$\frac{492 + 30}{492} \times 630.9, \text{ or } 669.1 \text{ B.T.U.}$$

The above calculation, like all giving accurate results, is somewhat tedious; a shorter and less correct one is as follows: Divide the figures found in the last column of

---

* Peabody's Steam Tables.

Table IV of the Appendix by 100, the result gives the heating value of these gases in B.T.U. per cubic centi-meter.* According to the volumetric analysis of the gas there are in 100 cc. the following:

  15.0 cc. illuminants,   25.3 cc. carbonic oxide;
  29.5 cc. methane,   27.9 cc. hydrogen;

the heating value is

$$15.0 \times 20.0 = 300.0 \text{ B.T.U.}$$
$$25.3 \times 3.41 = 86.3$$
$$25.9 \times 10.65 = 276.0$$
$$27.9 \times 3.45 = 96.3$$

$$758.6 \text{ B.T.U.}$$

the same as the gross heating value obtained by the other method. No correction is applied for the heat lost.

---

* Method followed in Prof. Paper, No. 48, U. S. Geol. Survey, Part III, p. 1005.

APPENDIX.

TABLE I.

TABLE SHOWING THE TENSION OF AQUEOUS VAPOR AND ALSO THE WEIGHT IN GRAMS CONTAINED IN A CUBIC METER OF AIR WHEN SATURATED.

From 5° to 30° C.

Temp.	Tension, mm.	Grams.	Temp.	Tension, mm.	Grams.	Temp.	Tension, mm.	Grams.
5	6.5	6.8	14	11.9	12.0	23	20.9	20.4
6	7.0	7.3	15	12.7	12.8	24	22.2	21.5
7	7.5	7.7	16	13.5	13.6	25	23.6	22.9
8	8.0	8.1	17	14.4	14.5	26	25.0	24.2
9	8.5	8.8	18	15.4	15.1	27	26.5	25.6
10	9.1	9.4	19	16.3	16.2	28	28.1	27.0
11	9.8	10.0	20	17.4	17.2	29	29.8	28.6
12	10.4	10.6	21	18.5	18.2	30	31.5	29.2
13	11.1	11.3	22	19.7	19.3			

TABLE II.

"VOLUMETRIC" SPECIFIC HEATS OF GASES.*

Air......................	0.019	"Illuminants".........	0.040
Carbon dioxide..........	0.027	Methane................	0.027
Carbonic oxide..........	0.019	Nitrogen...............	0.019
Hydrogen............	0.019	Oxygen................	0.019

The "volumetric" specific heat is the quantity of heat necessary to raise the temperature of one cubic foot of gas from 32° F. to 33° F.

---

* H. L. Payne, *Jour. Anal. and Applied Chem.*, 7, 233.

TABLE III.

THE VOLUME OF OXYGEN AND AIR NECESSARY TO BURN ONE CUBIC FOOT OF CERTAIN GASES, TOGETHER WITH THE VOLUME OF THE PRODUCTS OF COMBUSTION.

Name.	Formula.	Volume of Oxygen.	Volume* of Air.	Volume of Steam.	Volume of Carbon Dioxide.
Hydrogen	H_2	0.5	2.39	1	0
Carbonic oxide.	CO	0.5	2.39	0	1
Methane.......	CH_4	2.0	9.56	2	1
Ethane........	C_2H_6	3.5	16.73	3	2
Propane.......	C_3H_8	5.0	23.90	4	3
Butane........	C_4H_{10}	6.5	31.07	5	4
Pentane.......	C_5H_{12}	8.0	38.24	6	5
Hexane........	C_6H_{14}	9.5	45.41	7	6
Ethylene†.....	C_2H_4	3.0	14.34	2	2
Propylene‡....	C_3H_6	4.5	21.51	3	3
Benzene§......	C_6H_6	7.5	35.85	3	6

* Air being 20.92 per cent by volume, 4.78 volumes contain 1 volume of oxygen.

† The chief constituent of "illuminants," new name "ethene."

‡ New name "propene."

§ Often called benzol, not to be confounded with benz*i*ne.

TABLE IV.

CALORIFIC POWER OF VARIOUS GASES* IN BRITISH THERMAL UNITS
PER CUBIC FOOT.

Name.	Symbol.	60° initial. 328° final.§	32° initial. 32° final.
Hydrogen...............	H	263.2	345.4
Carbonic oxide.........	CO	306.9	341.2
Methane................	CH_4	853.0	1065.0
Illuminants†...........	....	1700.0	2000.0
Ethane.................	C_2H_6	....	1861.0
Propane................	C_3H_8	....	2657.0
Butane.................	C_4H_{10}	....	3441.0
Pentane................	C_5H_{12}	....	4255.0
Hexane‡................	C_6H_{14}	....	5017.0
Ethylene...............	C_2H_4	....	1674.0
Propylene..............	C_3H_6	....	2509.0
Benzene................	C_6H_6	....	4012.0

* H. L. Payne, *loc. cit.*

† Where the "illuminants" are derived chiefly from the decomposition of mineral oil.

‡ The chief constituent of the "gasolene" used in the gas
machines for *carburetting* air.

§ The temperature of steam at 100 lbs. absolute pressure.

TABLE V.

SHOWING THE WEIGHT OF A LITER AND SPECIFIC GRAVITY REFERRED
TO AIR, OF CERTAIN GASES AT 0° C. AND 760 MM.

Name of Gas.	Weight, Grams.	Specific Gravity.
Carbonic oxide.................	1.251	0.967
Carbon dioxide.................	1.966	1.519
Hydrogen......................	0.0896	0.069
Methane.......................	0.715	0.553
Nitrogen......................	1.255	0.970
Oxygen........................	1.430	1.105
Air...........................	1.294	1.000

TABLE VI.

SOLUBILITY OF VARIOUS GASES IN WATER.

One volume of water at 20° C. absorbs the following volumes of gas reduced to 0° C. and 760 mm. pressure.

Name of Gas.	Symbol.	Volumes.
Carbonic oxide..................	CO	0.023
Carbon dioxide..................	CO_2	0.901
Hydrogen.......................	H_2	0.019
Methane........................	CH_4	0.035
Nitrogen.......................	N_2	0.014
Oxygen.........................	O_2	0.028
Air............................	....	0.017

TABLE VII.

MELTING-POINTS OF VARIOUS METALS AND SALTS, FOR USE WITH APPARATUS FIG. II.

Alphabetically.		By Temperatures.	
Aluminium...........	660° C.*	Tin	233° C.‡
Antimony...........	432	Bismuth	268‡
†Barium chloride.....	922	Cadmium...........	320‡
Bismuth.............	268	Lead	334‡
Calcium fluoride......	902	Antimony...........	432‡
Cadmium............	320	Zinc	433‡
†Cadmium chloride...	541	Cadmium chloride...	541‡
Copper	1095	Aluminium..........	660*
Lead................	334	Potassium bromide..	722§
†Potassium bromide..	722	Sodium bromide.....	758§
†Potassium chloride..	800	Potassium chloride..	800§
†Sodium bromide.....	758	Sodium carbonate...	849§
†Sodium carbonate...	849	Calcium fluoride	902§
Tin	233	Barium chloride.....	922§
Zinc	433	Copper	1095*

* Holman, Proc. Am. Academy, **31**, 218 (1896).

† These salts must be dried at 105° C. to a constant **weight**.

‡ Carnelley, Melting- and Boiling-point Tables.

§ Meyer, Riddle and Lamb, Ber. d. deut. Chem. Gesellsch., **27**, 3140 (1894).

TABLE VIII.

GIVING THE NUMBER OF TIMES THE THEORETICAL QUANTITY OF AIR SUPPLIED, WITH VARIOUS GAS ANALYSES.*

$CO_2 + CO$.	$N = 70.$ $CO_2+O+CO=21$	$N = 80.$ $CO_2+O+CO=20$	$N = 81.$ $CO_2+O+CO=19$	$N = 82.$ $CO_2+O+CO=18$
21	1.00	....	....	....
20	1.05	1.00	....	....
19	1.10	1.05	....	....
18	1.17	1.10	1.00	....
17	1.23	1.16	1.05	1.00
16	1.31	1.23	1.10	1.05
15	1.40	1.31	1.16	1.10
14	1.50	1.39	1.23	1.16
13	1.61	1.49	1.30	1.22
12	1.75	1.60	1.39	1.30
11	1.91	1.73	1.48	1.38
10	2.10	1.89	1.59	1.47
9	2.33	2.07	1.72	1.58
8	2.62	2.29	1.87	1.70
7	3.00	2.57	2.04	1.85
6	3.50	2.92	2.26	2.02
5	4.20	3.39	2.52	2.23
4	5.25	4.05	2.86	2.48
3	7.00	5.00	3.30	2.79
2	10.50	6.53	3.89	3.20
1	21.00	9.43	4.76	3.76
			6.10	4.54

* Coxe, Proc. N. E. Cotton Manufacturers' Assoc., 1895.

TABLE IX.

COMPARISON OF METRIC AND ENGLISH SYSTEMS.

1 cubic inch	$= 16.39$ c.c.
1 cubic foot	$= 28.315$ liters.
1 Imperial gallon	$= 4.543$ "

---

1 lb. avoirdupois	$= 453.593$ grams.
1 calorie	$= 3.969$ B.T.U. (Röntgen).

INDEX.

SHORT-TITLE CATALOGUE

OF THE

PUBLICATIONS

OF

JOHN WILEY & SONS,

NEW YORK.

LONDON: CHAPMAN & HALL, LIMITED.

ARRANGED UNDER SUBJECTS.

Descriptive circulars sent on application. Books marked with an asterisk (*) are sold at *net* prices only. All books are bound in cloth unless otherwise stated.

AGRICULTURE—HORTICULTURE—FORESTRY.

Armsby's Principles of Animal Nutrition	8vo,	$4 00
Budd and Hansen's American Horticultural Manual:		
Part I. Propagation, Culture, and Improvement	12mo,	1 50
Part II. Systematic Pomology	12mo,	1 50
Elliott's Engineering for Land Drainage	12mo,	1 50
Practical Farm Drainage 2d Edition, Rewritten	12mo,	1 50
Graves's Forest Mensuration	8vo,	4 00
Green's Principles of American Forestry	12mo,	1 50
Grotenfelt's Principles of Modern Dairy Practice. (Woll)	12mo,	2 00
* Herrick's Denatured or Industrial Alcohol	8vo,	4 00
Kemp and Waugh's Landscape Gardening. New Edition, Rewritten. (In Preparation.)		
* McKay and Larsen's Principles and Practice of Butter-making	8vo,	1 50
Maynard's Landscape Gardening as Applied to Home Decoration	12mo,	1 50
Quaintance and Scott's Insects and Diseases of Fruits. (In Preparation).		
Sanderson's Insects Injurious to Staple Crops	12mo,	1 50
* Schwarz's Longleaf Pine in Virgin Forests	12mo,	1 25
Stockbridge's Rocks and Soils	8vo,	2 50
Winton's Microscopy of Vegetable Foods	8vo,	7 50
Woll's Handbook for Farmers and Dairymen	16mo,	1 50

ARCHITECTURE.

Baldwin's Steam Heating for Buildings	12mo,	2 50
Berg's Buildings and Structures of American Railroads	4to,	5 00
Birkmire's Architectural Iron and Steel	8vo,	3 50
Compound Riveted Girders as Applied in Buildings	8vo,	2 00
Planning and Construction of American Theatres	8vo,	3 00
Planning and Construction of High Office Buildings	8vo,	3 50
Skeleton Construction in Buildings	8vo,	3 00
Briggs's Modern American School Buildings	8vo,	4 00
Byrne's Inspection of Material and Wormanship Employed in Construction.	16mo,	3 00
Carpenter's Heating and Ventilating of Buildings	8vo,	4 00
* Corthell's Allowable Pressure on Deep Foundations	12mo,	1 25

Freitag's Architectural Engineering....................................8vo, 3 50
    Fireproofing of Steel Buildings...............................8vo, 2 50
French and Ives's Stereotomy....................................8vo, 2 50
Gerhard's Guide to Sanitary House-Inspection....................16mo, 1 00
*   Modern Baths and Bath Houses..............................8vo, 3 00
    Sanitation of Public Buildings..............................12mo, 1 50
    Theatre Fires and Panics...................................12mo, 1 50
Holley and Ladd's Analysis of Mixed Paints, Color Pigments, and Varnishes
                                             Large 12mo, 2 50
Johnson's Statics by Algebraic and Graphic Methods.................8vo, 2 00
Kellaway's How to Lay Out Suburban Home Grounds................8vo, 2 00
Kidder's Architects' and Builders' Pocket-book. ....16mo, mor. 5 00
Maire's Modern Pigments and their Vehicles 12mo, 2 00
Merrill's Non-metallic Minerals: Their Occurrence and Uses..........8vo, 4 00
    Stones for Building and Decoration...........................8vo, 5 00
Monckton's Stair-building..4to, 4 00
Patton's Practical Treatise on Foundations.........................8vo, 5 00
Peabody's Naval Architecture...................................8vo, 7 50
Rice's Concrete-block Manufacture 8vo, 2 00
Richey's Handbook for Superintendents of Construction..........16mo, mor. 4 00
    * Building Mechanics' Ready Reference Book:
        * Building Foreman's Pocket Book and Ready Reference. (In
            Press.)
        * Carpenters' and Woodworkers' Edition..............16mo, mor. 1 50
        * Cement Workers and Plasterer's Edition...........16mo, mor. 1 50
        * Plumbers', Steam-Filters', and Tinners' Edition....16mo, mor. 1 50
        * Stone- and Brick-masons' Edition..................16mo, mor. 1 50
Sabin's House Painting 12mo, 1 00
    Industrial and Artistic Technology of Paints and Varnish..........8vo, 3 00
Siebert and Biggin's Modern Stone-cutting and Masonry..............8vo, 1 50
Snow's Principal Species of Wood................................8vo, 3 50
Towne's Locks and Builders' Hardware.......................18mo, mor. 3 00
Wait's Engineering and Architectural Jurisprudence 8vo, 6 00
                                        Sheep, 6 50
    Law of Contracts. 8vo, 3 00
    Law of Operations Preliminary to Construction in Engineering and Archi-
        tecture..8vo, 5 00
                                        Sheep, 5 50
Wilson's Air Conditioning.......................................12mo, 1 50
Worcester and Atkinson's Small Hospitals, Establishment and Maintenance,
    Suggestions for Hospital Architecture, with Plans for a Small Hospital.
                                        12mo, 1 25

ARMY AND NAVY.

Bernadou's Smokeless Powder, Nitro-cellulose, and the Theory of the Cellulose
    Molecule..12mo, 2 50
Chase's Art of Pattern Making..................................12mo, 2 50
    Screw Propellers and Marine Propulsion.......................8vo, 3 00
* Cloke's Enlisted Specialist's Examiner...........................8vo, 2 00
    Gunner's Examiner 8vo, 1 50
Craig's Azimuth...4to, 3 50
Crehore and Squier's Polarizing Photo-chronograph.................8vo, 3 00
* Davis's Elements of Law......................................8vo, 2 50
*   Treatise on the Military Law of United States.................8vo, 7 00
                                        Sheep, 7 50
De Brack's Cavalry Outpost Duties. (Carr)....................24mo, mor. 2 00
* Dudley's Military Law and the Procedure of Courts-martial...Large 12mo, 2 50
Durand's Resistance and Propulsion of Ships......................8vo, 5 00

ASSAYING.

ASTRONOMY.

CHEMISTRY.

* Abderhalden's Physiological Chemistry in Thirty Lectures.   (Hall and Defren)
8vo, 5 00
* Abegg's Theory of Electrolytic Dissociation. (von Ende)........... 12mo,   1 25
Alexeyeff's General Principles of Organic Syntheses.   (Matthews)........8vo,   3 00
Allen's Tables for Iron Analysis.............................8vo,   3 00
Arnold's Compendium of Chemistry.   (Mandel)............. Large 12mo,   3 50
Association of State and National Food and Dairy Departments, Hartford,
  Meeting, 1906 ....8vo,   3 00
  Jamestown Meeting, 1907 ....8vo,   3 00
Austen's Notes for Chemical Students 12mo,   1 50
Baskerville's Chemical Elements. (In Preparation.)
Bernadou's Smokeless Powder.—Nitro-cellulose, and Theory of the Cellulose
  Molecule....................................12mo,   2 50
Bilts's Chemical Preparations. (Hall and Blanchard). (In Press.)
* Blanchard's Synthetic Inorganic Chemistry.......................12mo,   1 00
* Browning's Introduction to the Rarer Elements....................8vo,   1 50
Brush and Penfield's Manual of Determinative Mineralogy............8vo,   4 00
* Claassen's Beet-sugar Manufacture. (Hall and Rolfe).............8vo,   3 00
Classen's Quantitative Chemical Analysis by Electrolysis. (Boltwood)..8vo,   3 00
Cohn's Indicators and Test-papers.........................12mo,   2 00
  Tests and Reagents...................................8vo,   3 00
* Danneel's Electrochemistry. (Merriam).....................12mo,   1 25
Dannerth's Methods of Textile Chemistry.........................12mo,   2 00
Duhem's Thermodynamics and Chemistry. (Burgess)...............8vo,   4 00
Eakle's Mineral Tables for the Determination of Minerals by their Physical
  Properties....................................8vo,   1 25
Eissler's Modern High Explosives.............................8vo,   4 00
Effront's Enzymes and their Applications. (Prescott)...............8vo,   3 00
Erdmann's Introduction to Chemical Preparations. (Dunlap).........12mo,   1 25
* Fischer's Physiology of Alimentation . ...Large 12mo,   2 00
Fletcher's Practical Instructions in Quantitative Assaying with the Blowpipe.
12mo, mor.   1 50
Fowler's Sewage Works Analyses............................12mo,   2 00
Fresenius's Manual of Qualitative Chemical Analysis. (Wells)........8vo,   5 00
  Manual of Qualitative Chemical Analysis. Part I. Descriptive. (Wells) 8vo,   3 00
  Quantitative Chemical Analysis. (Cohn) 2 vols...............8vo, 12 50
      When Sold Separately, Vol. I, $6.   Vol. II, $8.
Fuertes's Water and Public Health...........................12mo,   1 50
Furman's Manual of Practical Assaying.........................8vo,   3 00
* Getman's Exercises in Physical Chemistry.....................12mo,   2 00
Gill's Gas and Fuel Analysis for Engineers.....................12mo,   1 25
* Gooch and Browning's Outlines of Qualitative Chemical Analysis.
Large 12mo,   1 25
Grotenfelt's Principles of Modern Dairy Practice. (Woll)...........12mo,   2 00
Groth's Introduction to Chemical Crystallography (Marshall) 12mo.   1 25
Hammarsten's Text-book of Physiological Chemistry. (Mandel).......8vo,   4 00
Hanausek's Microscopy of Technical Products. (Winton)..............8vo,   5 00
* Haskins and Macleod's Organic Chemistry 12mo,   2 00
Helm's Principles of Mathematical Chemistry. (Morgan)...........12mo,   1 50
Hering's Ready Reference Tables (Conversion Factors)..........16mo, mor.   2 50
* Herrick's Denatured or Industrial Alcohol 8vo   4 00
Hinds's Inorganic Chemistry.................................8vo,   3 00
*   Laboratory Manual for Students 12mo,   1 00
* Holleman's Laboratory Manual of Organic Chemistry for Beginners.
  (Walker).....................................12mo,   1 00
  Text-book of Inorganic Chemistry. (Cooper)...............8vo,   2 50
  Text-book of Organic Chemistry. (Walker and Mott)...........8vo,   2 50

4

olley and Ladd's Analysis of Mixed Paints, Color Pigments, and Varnishes.

CIVIL ENGINEERING.

BRIDGES AND ROOFS. HYDRAULICS. MATERIALS OF ENGINEERING. RAILWAY ENGINEERING.

Hayford's Text-book of Geodetic Astronomy.................................8vo, 3 00
Hering's Ready Reference Tables. (Conversion Factors)........16mo, mor. 2 50
Howe's Retaining Walls for Earth..............................12mo, 1 25
* Ives's Adjustments of the Engineer's Transit and Level..........16mo, Bds. 25
Ives and Hilts's Problems in Surveying.....................16mo, mor. 1 50
Johnson's (J. B.) Theory and Practice of Surveying.............Small 8vo, 4 00
Johnson's (L. J.) Statics by Algebraic and Graphic Methods............8vo, 2 00
Kinnicutt, Winslow and Pratt's Purification of Sewage. (In Preparation.)
Laplace's Philosophical Essay on Probabilities. (Truscott and Emory)
12mo, 2 00
Mahan's Descriptive Geometry.....................................12mo, 1 50
    Treatise on Civil Engineering. (1873.) (Wood)................8vo, 5 00
Merriman's Elements of Precise Surveying and Geodesy...............8vo, 2 50
Merriman and Brooks's Handbook for Surveyors...............16mo, mor. 2 00
Nugent's Plane Surveying...8vo, 3 50
Ogden's Sewer Construction.......................................8vo, 3 00
    Sewer Design..12mo, 2 00
Parsons's Disposal of Municipal Refuse............................8vo, 2 00
Patton's Treatise on Civil Engineering...............8vo, half leather, 7 50
Reed's Topographical Drawing and Sketching 4to, 5 00
Rideal's Sewage and the Bacterial Purification of Sewage............8vo, 4 00
Riemer's Shaft-sinking under Difficult Conditions. (Corning and Peele)...8vo, 3 00
Siebert and Biggin's Modern Stone-cutting and Masonry..............8vo, 1 50
Smith's Manual of Topographical Drawing. (McMillan)...............8vo, 2 50
Soper's Air and Ventilation of Subways ....Large 12mo, 2 50
Tracy's Plane Surveying....................................16mo, mor. 3 00
* Trautwine's Civil Engineer's Pocket-book.................16mo, mor. 5 00
Venable's Garbage Crematories in America.........................8vo, 2 00
    Methods and Devices for Bacterial Treatment of Sewage........8vo, 3 00
Wait's Engineering and Architectural Jurisprudence................8vo, 6 00
                                                            Sheep, 6 50
    Law of Contracts..8vo, 3 00
    Law of Operations Preliminary to Construction in Engineering and Archi-
    tecture...8vo, 5 00
                                                            Sheep, 5 50
Warren's Stereotomy—Problems in Stone-cutting.....................8vo, 2 50
* Waterbury's Vest-Pocket Hand-book of Mathematics for Engineers.
                                        2⅞ × 5⅝ inches, mor. 1 00
Webb's Problems in the Use and Adjustment of Engineering Instruments.
                                                        16mo, mor. 1 25
Wilson's (H. N.) Topographic Surveying 8vo, 3 50
Wilson's (W. L.) Elements of Railroad Track and Construction.......12mo, 2 00

BRIDGES AND ROOFS.

Boller's Practical Treatise on the Construction of Iron Highway Bridges..8vo, 2 00
Burr and Falk's Design and Construction of Metallic Bridges 8vo, 5 00
    Influence Lines for Bridge and Roof Computations..............8vo, 3 00
Du Bois's Mechanics of Engineering. Vol. II.................Small 4to, 10 00
Foster's Treatise on Wooden Trestle Bridges......................4to, 5 00
Fowler's Ordinary Foundations...................................8vo, 3 50
French and Ives's Stereotomy....................................8vo, 2 50
Greene's Arches in Wood, Iron, and Stone.........................8vo, 2 50
    Bridge Trusses..8vo, 2 50
    Roof Trusses..8vo, 1 25
Grimm's Secondary Stresses in Bridge Trusses.....................8vo, 2 50
Heller's Stresses in Structures and the Accompanying Deformations.....8vo, 3 00
Howe's Design of Simple Roof-trusses in Wood and Steel............8vo, 2 00
    Symmetrical Masonry Arches...............................8vo, 2 50
    Treatise on Arches..8vo, 4 00

Johnson, Bryan, and Turneaure's Theory and Practice in the Designing of
        Modern Framed Structures...Small 4to, 10 00
Merriman and Jacoby's Text-book on Roofs and Bridges:
    Part I.   Stresses in Simple Trusses.............................8vo, 2 50
    Part II. Graphic Statics..8vo, 2 50
    Part III. Bridge Design..8vo, 2 50
    Part IV. Higher Structures......................................8vo, 2 50
Morison's Memphis Bridge.............................Oblong 4to, 10 00
Sondericker's Graphic Statics, with Applications to Trusses, Beams, and Arches.
                                                              8vo, 2 00
Waddell's De Pontibus, Pocket-book for Bridge Engineers...... 16mo, mor, 2 00
*   Specifications for Steel Bridges............................12mo,   50
Waddell and Harrington's Bridge Engineering. (In Preparation.)
Wright's Designing of Draw-spans. Two parts in one volume.........8vo, 3 50

HYDRAULICS.

Barnes's Ice Formation...8vo, 3 00
Bazin's Experiments upon the Contraction of the Liquid Vein Issuing from
        an Orifice. (Trautwine)....................................8vo, 2 00
Bovey's Treatise on Hydraulics......................................8vo, 5 00
Church's Diagrams of Mean Velocity of Water in Open Channels.
                                                    Oblong 4to, paper, 1 50
    Hydraulic Motors..8vo, 2 00
    Mechanics of Engineering..8vo, 6 00
Coffin's Graphical Solution of Hydraulic Problems.............16mo, mor, 2 50
Flather's Dynamometers, and the Measurement of Power............12mo, 3 00
Folwell's Water-supply Engineering..................................8vo, 4 00
Frizell's Water-power...8vo, 5 00
Fuertes's Water and Public Health..................................12mo, 1 50
    Water-filtration Works..12mo, 2 50
Ganguillet and Kutter's General Formula for the Uniform Flow of Water in
        Rivers and Other Channels. (Hering and Trautwine)........ 8vo, 4 00
Hazen's Clean Water and How to Get It....................Large 12mo, 1 50
    Filtration of Public Water-supplies.............................8vo, 3 00
Hazlehurst's Towers and Tanks for Water-works.......................8vo, 2 50
Herschel's 115 Experiments on the Carrying Capacity of Large, Riveted, Metal
        Conduits..8vo, 2 00
Hoyt and Grover's River Discharge...................................8vo, 2 00
Hubbard and Kiersted's Water-works Management and Maintenance.....8vo, 4 00
* Lyndon's Development and Electrical Distribution of Water Power....8vo, 3 00
Mason's Water-supply. (Considered Principally from a Sanitary Standpoint.)
                                                              8vo, 4 00
Merriman's Treatise on Hydraulics..................................8vo, 5 00
* Michie's Elements of Analytical Mechanics.........................8vo, 4 00
* Molitor's Hydraulics of Rivers, Weirs and Sluices.................8vo, 2 00
Richards's Laboratory Notes on Industrial Water Analysis. (In Press.)
Schuyler's Reservoirs for Irrigation, Water-power, and Domestic Water-
        supply...Large 8vo, 5 00
* Thomas and Watt's Improvement of Rivers...........................4to, 6 00
Turneaure and Russell's Public Water-supplies.......................8vo, 5 00
Wegmann's Design and Construction of Dams. 5th Ed., enlarged.....4to, 6 00
    Water-supply of the City of New York from 1658 to 1895..........4to, 10 00
Whipple's Value of Pure Water............................Large 12mo, 1 00
Williams and Hazen's Hydraulic Tables...............................8vo, 1 50
Wilson's Irrigation Engineering.............................Small 8vo, 4 00
Wolff's Windmill as a Prime Mover...................................8vo, 3 00
Wood's Elements of Analytical Mechanics............................8vo, 3 00
    Turbines..8vo, 2 50

8

MATERIALS OF ENGINEERING.

Wood's (De V.) Treatise on the Resistance of Materials, and an Appendix on
the Preservation of Timber. 8vo, 2 00
Wood's (M. P.) Rustless Coatings: Corrosion and Electrolysis of Iron and
Steel. 8vo, 4 00

RAILWAY ENGINEERING.

Andrews's Handbook for Street Railway Engineers ....3x5 inches, mor. 1 25
Berg's Buildings and Structures of American Railroads 4to, 5 00
Brooks's Handbook of Street Railroad Location. ....16mo, mor. 1 50
Butt's Civil Engineer's Field-book. ....16mo, mor. 2 50
Crandall's Railway and Other Earthwork Tables. 8vo, 1 50
Transition Curve. ....16mo, mor. 1 50
* Crockett's Methods for Earthwork Computations. 8vo, 1 50
Dawson's "Engineering" and Electric Traction Pocket-book. .....16mo, mor. 5 00
Dredge's History of the Pennsylvania Railroad: (1879). ....Paper, 5 00
Fisher's Table of Cubic Yards. ....Cardboard,   25
Godwin's Railroad Engineers' Field-book and Explorers' Guide... 16mo, mor. 2 50
Hudson's Tables for Calculating the Cubic Contents of Excavations and Em-
bankments. 8vo, 1 00
Ives and Hilts's Problems in Surveying, Railroad Surveying and Geodesy
16mo, mor. 1 50
Molitor and Beard's Manual for Resident Engineers. 16mo, 1 00
Nagle's Field Manual for Railroad Engineers. ....16mo, mor. 3 00
Philbrick's Field Manual for Engineers. ....16mo, mor. 3 00
Raymond's Railroad Engineering. 3 volumes.
Vol. I. Railroad Field Geometry. (In Preparation.)
Vol. II. Elements of Railroad Engineering. 8vo, 3 50
Vol. III. Railroad Engineer's Field Book. (In Preparation.)
Searles's Field Engineering. ....16mo, mor. 3 00
Railroad Spiral. ....16mo, mor. 1 50
Taylor's Prismoidal Formulæ and Earthwork. 8vo, 1 50
*Trautwine's Field Practice of Laying Out Circular Curves for Railroads.
12mo. mor. 2 50
* Method of Calculating the Cubic Contents of Excavations and Embank-
ments by the Aid of Diagrams. 8vo, 2 00
Webb's Economics of Railroad Construction. ....Large 12mo, 2 50
Railroad Construction. ....16mo, mor. 5 00
Wellington's Economic Theory of the Location of Railways. ......Small 8vo, 5 00

DRAWING.

Barr's Kinematics of Machinery. 8vo, 2 50
* Bartlett's Mechanical Drawing. 8vo, 3 00
*   "   "   "   Abridged Ed. 8vo, 1 50
Coolidge's Manual of Drawing. ....8vo, paper, 1 00
Coolidge and Freeman's Elements of General Drafting for Mechanical Engi-
neers. ....Oblong 4to, 2 50
Durley's Kinematics of Machines. 8vo, 4 00
Emch's Introduction to Projective Geometry and its Applications. 8vo, 2 50
Hill's Text-book on Shades and Shadows, and Perspective. 8vo, 2 00
Jamison's Advanced Mechanical Drawing. 8vo, 2 00
Elements of Mechanical Drawing. 8vo, 2 50
Jones's Machine Design:
Part I. Kinematics of Machinery. 8vo, 1 50
Part II. Form, Strength, and Proportions of Parts. 8vo, 3 00
MacCord's Elements of Descriptive Geometry. 8vo, 3 00
Kinematics; or, Practical Mechanism. 8vo, 5 00
Mechanical Drawing. 4to, 4 00
Velocity Diagrams. 8vo, 1 50

McLeod's Descriptive Geometry............................Large 12mo, 1 50
* Mahan's Descriptive Geometry and Stone-cutting...................8vo, 1 50
    Industrial Drawing. (Thompson)............................8vo, 3 50
Moyer's Descriptive Geometry.................................8vo, 2 00
Reed's Topographical Drawing and Sketching......................4to, 5 00
Reid's Course in Mechanical Drawing...........................8vo, 2 00
    Text-book of Mechanical Drawing and Elementary Machine Design. 8vo, 3 00
Robinson's Principles of Mechanism............................8vo, 3 00
Schwamb and Merrill's Elements of Mechanism....................8vo, 3 00
Smith's (R. S.) Manual of Topographical Drawing. (McMillan).......8vo, 2 50
Smith (A. W.) and Marx's Machine Design........................8vo, 3 00
* Titsworth's Elements of Mechanical Drawing.............Oblong 8vo, 1 25
Warren's Drafting Instruments and Operations...................12mo, 1 25
    Elements of Descriptive Geometry, Shadows, and Perspective........8vo, 3 50
    Elements of Machine Construction and Drawing..................8vo, 7 50
    Elements of Plane and Solid Free-hand Geometrical Drawing.....12mo, 1 00
    General Problems of Shades and Shadows.....................8vo, 3 00
    Manual of Elementary Problems in the Linear Perspective of Form and
      Shadow...12mo, 1 00
    Manual of Elementary Projection Drawing.....................12mo, 1 50
    Plane Problems in Elementary Geometry......................12mo, 1 25
    Problems, Theorems, and Examples in Descriptive Geometry.......8vo, 2 50
Weisbach's Kinematics and Power of Transmission. (Hermann and
    Klein)...8vo, 5 00
Wilson's (H. M.) Topographic Surveying.........................8vo, 3 50
Wilson's (V. T.) Free-hand Lettering...........................8vo, 1 00
    Free-hand Perspective....................................8vo, 2 50
Woolf's Elementary Course in Descriptive Geometry............Large 8vo, 3 00

ELECTRICITY AND PHYSICS.

* Abegg's Theory of Electrolytic Dissociation. (von Ende).........12mo, 1 25
Andrews's Hand-Book for Street Railway Engineering ....3×5 inches, mor. 1 25
Anthony and Brackett's Text-book of Physics. (Magie).......Large 12mo, 3 00
Anthony's Theory of Electrical Measurements. (Ball).............12mo, 1 00
Benjamin's History of Electricity.............................8vo, 3 00
    Voltaic Cell...8vo, 3 00
Betts's Lead Refining and Electrolysis.........................8vo, 4 00
Classen's Quantitative Chemical Analysis by Electrolysis. (Boltwood)..8vo, 3 00
* Collins's Manual of Wireless Telegraphy.....................12mo, 1 50
                                         Mor. 2 00
Crehore and Squier's Polarizing Photo-chronograph..................8vo, 3 00
* Danneel's Electrochemistry. (Merriam)......................12mo, 1 25
Dawson's "Engineering" and Electric Traction Pocket-book ....16mo, mor. 5 00
Dolezalek's Theory of the Lead Accumulator (Storage Battery). (von Ende)
                                        12mo, 2 50
Duhem's Thermodynamics and Chemistry. (Burgess)................8vo, 4 00
Flather's Dynamometers, and the Measurement of Power............12mo, 3 00
Gilbert's De Magnete. (Mottelay).............................8vo, 2 50
* Hanchett's Alternating Currents............................12mo, 1 00
Hering's Ready Reference Tables (Conversion Factors)..........16mo, mor. 2 50
* Hobart and Ellis's High-speed Dynamo Electric Machinery 8vo, 6 00
Holman's Precision of Measurements............................8vo, 2 00
    Telescopic Mirror-scale Method, Adjustments, and Tests....Large 8vo, 75
* Karapetoff's Experimental Electrical Engineering.................8vo, 6 00
Kinzbrunner's Testing of Continuous-current Machines.............8vo, 2 00
Landauer's Spectrum Analysis. (Tingle).......................8vo, 3 00
Le Chatelier's High-temperature Measurements. (Boudouard—Burgess)..12mo, 3 00
Löb's Electrochemistry of Organic Compounds. (Lorenz).............8vo, 3 00
* Lyndon's Development and Electrical Distribution of Water Power 8vo, 3 00

11

* Lyons's Treatise on Electromagnetic Phenomena. Vols. I. and II. 8vo, each, 6 00
* Michie's Elements of Wave Motion Relating to Sound and Light. 8vo, 4 00
Morgan's Outline of the Theory of Solution and its Results...........12mo, 1 00
*   Physical Chemistry for Electrical Engineers....................12mo, 1 50
Niaudet's Elementary Treatise on Electric Batteries. (Fishback)....12mo, 2 50
* Norris's Introduction to the Study of Electrical Engineering.......8vo, 2 50
* Parshall and Hobart's Electric Machine Design.............4to, half mor. 12 50
Reagan's Locomotives: Simple, Compound, and Electric. New Edition.
                                                Large 12mo, 3 50
* Rosenberg's Electrical Engineering. (Haldane Gee—Kinzbrunner). ...8vo, 2 00
Ryan, Norris, and Hoxie's Electrical Machinery. Vol. I..............8vo, 2 50
Schapper's Laboratory Guide for Students in Physical Chemistry 12mo, 1 00
* Tillman's Elementary Lessons in Heat.............................8vo, 1 50
Tory and Pitcher's Manual of Laboratory Physics............Large 12mo, 2 00
Ulke's Modern Electrolytic Copper Refining.......................8vo, 3 00

LAW.

Brennan's Handbook: A Compendium of Useful Legal Information for
        Business Men.......................................16mo, mor. 5.00
* Davis's Elements of Law....................................8vo, 2 50
*   Treatise on the Military Law of United States...................8vo, 7 00
*   Sheep, 7 50
* Dudley's Military Law and the Procedure of Courts-martial ....Large 12mo, 2 50
Manual for Courts-martial...................................16mo, mor. 1 50
Wait's Engineering and Architectural Jurisprudence..................8vo, 6 00
                                                       Sheep, 6 50
    Law of Contracts..8vo, 3 00
    Law of Operations Preliminary to Construction in Engineering and Archi-
        tecture..8vo, 5 00
                                                       Sheep, 5 50

MATHEMATICS.

Baker's Elliptic Functions 8vo, 1 50
Briggs's Elements of Plane Analytic Geometry. (Bôcher)............12mo, 1 00
* Buchanan's Plane and Spherical Trigonometry.....................8vo, 1 00
Byerley's Harmonic Functions..................................8vo, 1 00
Chandler's Elements of the Infinitesimal Calculus...................12mo, 2 00
Coffin's Vector Analysis.   (In Press.)
Compton's Manual of Logarithmic Computations 12mo, 1 50
* Dickson's College Algebra..............................Large 12mo, 1 50
*   Introduction to the Theory of Algebraic Equations ....Large 12mo, 1 25
Emch's Introduction to Projective Geometry and its Applications 8vo, 2 50
Fiske's Functions of a Complex Variable.........................8vo, 1 00
Halsted's Elementary Synthetic Geometry 8vo, 1 50
    Elements of Geometry 8vo, 1 75
*   Rational Geometry......................................12mo, 1 50
Hyde's Grassmann's Space Analysis............................8vo, 1 00
* Jonnson's (J. B.) Three-place Logarithmic Tables: Vest-pocket size, paper, 15
                                                100 copies, 5 00
*   Mounted on heavy cardboard, 8 × 10 inches, 25
                                                10 copies, 2 00
Johnson's (W. W.) Abridged Editions of Differential and Integral Calculus
                                                Large 12mo, 1 vol. 2 50
    Curve Tracing in Cartesian Co-ordinates 12mo, 1 00
    Differential Equations......................................8vo, 1 00
    Elementary Treatise on Differential Calculus ....Large 12mo, 1 50
    Elementary Treatise on the Integral Calculus ....Large 12mo, 1 50
*   Theoretical Mechanics...................................12mo, 3 00
    Theory of Errors and the Method of Least Squares...........12mo, 1 50
    Treatise on Differential Calculus........................Large 12mo, 3 00

Johnson's Treatise on the Integral Calculus...................Large 12mo, 3 00
      Treatise on Ordinary and Partial Differential Equations..Large 12mo, 3 50
Karapetoff's Engineering Applications of Higher Mathematics. (In Preparation.)

Laplace's Philosophical Essay on Probabilities. (Truscott and Emory)..12mo, 2 00
* Ludlow and Bass's Elements of Trigonometry and Logarithmic and Other
    Tables..8vo, 3 00
    Trigonometry and Tables published separately..................Each, 2 00
* Ludlow's Logarithmic and Trigonometric Tables...................8vo, 1 00
Macfarlane's Vector Analysis and Quaternions......................8vo, 1 00
McMahon's Hyperbolic Functions..................................8vo, 1 00
Manning's Irrational Numbers and their Representation by Sequences and
    Series..12mo, 1 25
Mathematical Monographs. Edited by Mansfield Merriman and Robert
    S. Woodward..Octavo, each 1 00
    No. 1. History of Modern Mathematics, by David Eugene Smith.
    No. 2. Synthetic Projective Geometry, by George Bruce Halsted.
    No. 3. Determinants, by Laenas Gifford Weld. No. 4. Hyperbolic Functions, by James McMahon. No. 5. Harmonic Functions, by William E. Byerly. No. 6. Grassmann's Space Analysis, by Edward W. Hyde. No. 7. Probability and Theory of Errors, by Robert S. Woodward. No. 8. Vector Analysis and Quaternions, by Alexander Macfarlane. No. 9. Differential Equations, by William Woolsey Johnson. No. 10. The Solution of Equations, by Mansfield Merriman. No. 11. Functions of a Complex Variable, by Thomas S. Fiske.
Maurer's Technical Mechanics.....................................8vo, 4 00
Merriman's Method of Least Squares..............................8vo, 2 00
    Solution of Equations.......................................8vo, 1 00
Rice and Johnson's Differential and Integral Calculus. 2 vols. in one.
                                    Large 12mo, 1 50
    Elementary Treatise on the Differential Calculus..........Large 12mo, 3 00
Smith's History of Modern Mathematics.............................8vo, 1 00
* Veblen and Lennes's Introduction to the Real Infinitesimal Analysis of One
    Variable..8vo, 2 00
* Waterbury's Vest Pocket Hand-Book of Mathematics for Engineers.
                          $2\frac{7}{8} \times 5\frac{3}{8}$ inches, mor. 1 00
Weld's Determinations..8vo, 1 00
Wood's Elements of Co-ordinate Geometry..........................8vo, 2 00
Woodward's Probability and Theory of Errors.......................8vo, 1 00

MECHANICAL ENGINEERING.

MATERIALS OF ENGINEERING, STEAM-ENGINES AND BOILERS.

Bacon's Forge Practice..12mo, 1 50
Baldwin's Steam Heating for Buildings............................12mo, 2 50
Barr's Kinematics of Machinery...................................8vo, 2 50
* Bartlett's Mechanical Drawing..................................8vo, 3 00
*   "   "   "   Abridged Ed......................8vo, 1 50
Benjamin's Wrinkles and Recipes.................................12mo, 2 00
* Burr's Ancient and Modern Engineering and the Isthmian Canal......8vo, 3 50
Carpenter's Experimental Engineering............................8vo, 6 00
    Heating and Ventilating Buildings...........................8vo, 4 00
Clerk's Gas and Oil Engine...................................Large 12mo, 4 00
Compton's First Lessons in Metal Working........................12mo, 1 50
Compton and De Groodt's Speed Lathe............................12mo, 1 50
Coolidge's Manual of Drawing...............................8vo, paper, 1 00
Coolidge and Freeman's Elements of General Drafting for Mechanical Engineers..Oblong 4to, 2 50

13

MATERIALS OF ENGINEERING

Holley and Ladd's Analysis of Mixed Paints, Color Pigments, and Varnishes.
Large 12mo, 2 50
Johnson's Materials of Construction..................................8vo, 6 00
Keep's Cast Iron...8vo, 2 50
Lanza's Applied Mechanics..8vo, 7 50
Maire's Modern Pigments and their Vehicles......................12mo, 2 00
Martens's Handbook on Testing Materials. (Henning)..............8vo, 7 50
Maurer's Technical Mechanics.......................................8vo, 4 00
Merriman's Mechanics of Materials..................................8vo, 5 00
*   Strength of Materials.......................................12mo, 1 00
Metcaif's Steel. A Manual for Steel-users.........................12mo, 2 00
Sabin's Industrial and Artistic Technology of Paints and Varnish........8vo, 3 00
Smith's Materials of Machines.....................................12mo, 1 00
Thurston's Materials of Engineering....................3 vols., 8vo, 8 00
    Part I.   Non-metallic Materials of Engineering and Metallurgy...8vo, 2 00
    Part II. Iron and Steel.......................................8vo, 3 50
    Part III. A Treatise on Brasses, Bronzes, and Other Alloys and their
              Constituents...8vo, 2 50
Wood's (De V.) Elements of Analytical Mechanics....................8vo, 3 00
    Treatise on the Resistance of Materials and an Appendix on the
    Preservation of Timber 8vo, 2 00
Wood's (M. P.) Rustless Coatings: Corrosion and Electrolysis of Iron and
    Steel...8vo, 4 00

STEAM-ENGINES AND BOILERS.

Berry's Temperature-entropy Diagram..............................12mo, 1 25
Carnot's Reflections on the Motive Power of Heat. (Thurston)......12mo, 1 50
Chase's Art of Pattern Making....................................12mo, 2 50
Creighton's Steam-engine and other Heat-motors. 8vo, 5 00
Dawson's " Engineering " and Electric Traction Pocket-book.....16mo, mor. 5 00
Ford's Boiler Making for Boiler Makers............................18mo, 1 00
*Gebhardt's Steam Power Plant Engineering 8vo, 6 00
Goss's Locomotive Performance......................................8vo, 5 00
Hemenway's Indicator Practice and Steam-engine Economy...........12mo, 2 00
Hutton's Heat and Heat-engines....................................8vo, 5 00
    Mechanical Engineering of Power Plants........................8vo, 5 00
Kent's Steam boiler Economy.......................................8vo, 4 00
Kneass's Practice and Theory of the Injector......................8vo, 1 50
MacCord's Slide-valves..8vo, 2 00
Meyer's Modern Locomotive Construction............................4to, 10 00
Moyer's Steam Turbines.   (In Press.)
Peabody's Manual of the Steam-engine Indicator...................12mo, 1 50
    Tables of the Properties of Saturated Steam and Other Vapors.....8vo, 1 00
    Thermodynamics of the Steam-engine and Other Heat-engines......8vo, 5 00
    Valve-gears for Steam-engines.................................8vo, 2 50
Peabody and Miller's Steam-boilers................................8vo, 4 00
Pray's Twenty Years with the Indicator......................Large 8vo, 2 50
Pupin's Thermodynamics of Reversible Cycles in Gases and Saturated Vapors.
    (Osterberg)...12mo, 1 25
Reagan's Locomotives: Simple, Compound, and Electric. New Edition.
Large 12mo, 3 50
Sinclair's Locomotive Engine Running and Management..............12mo, 2 00
Smart's Handbook of Engineering Laboratory Practice..............12mo, 2 50
Snow's Steam-boiler Practice......................................8vo, 3 00
Spangler's Notes on Thermodynamics...............................12mo, 1 00
    Valve-gears...8vo, 2 50
Spangler, Greene, and Marshall's Elements of Steam-engineering 8vo, 3 00
Thomas's Steam-turbines...8vo, 4 00

Thurston's Handbook of Engine and Boiler Trials, and the Use of the Indicator and the Prony Brake..........................8vo, 5 00
    Handy Tables.......................................8vo, 1 50
    Manual of Steam-boilers, their designs, Construction, and Operation..8vo, 5 00
Thurston's Manual of the Steam-engine......................2 vols., 8vo, 10 00
    Part I. History, Structure, and Theory....................8vo, 6 00
    Part II. Design, Construction, and Operation.................8vo, 6 00
    Steam-boiler Explosions in Theory and in Practice............12mo, 1 50
Wehrenfenning's Analysis and Softening of Boiler Feed-water (Patterson) 8vo, 4 00
Weisbach's Heat, Steam, and Steam-engines. (Du Bois).............8vo, 5 00
Whitham's Steam-engine Design.................................8vo, 5 00
Wood's Thermodynamics, Heat Motors, and Refrigerating Machines...8vo, 4 00

MECHANICS PURE AND APPLIED.

Church's Mechanics of Engineering............................8vo, 6 00
    Notes and Examples in Mechanics..........................8vo, 2 00
Dana's Text-book of Elementary Mechanics for Colleges and Schools..12mo, 1 50
Du Bois's Elementary Principles of Mechanics:
    Vol. I. Kinematics......................................8vo, 3 50
    Vol. II. Statics...8vo, 4 00
    Mechanics of Engineering. Vol. I.....................Small 4to, 7 50
                  Vol. II.......................Small 4to, 10 00
* Greene's Structural Mechanics................................8vo, 2 50
James's Kinematics of a Point and the Rational Mechanics of a Particle.
                                         Large 12mo, 2 00
* Johnson's (W. W.) Theoretical Mechanics....................12mo, 3 00
Lanza's Applied Mechanics.....................................8vo, 7 50
* Martin's Text Book on Mechanics, Vol. I, Statics................12mo, 1 25
*   Vol. 2, Kinematics and Kinetics..12mo, 1 50
Maurer's Technical Mechanics..................................8vo, 4 00
* Merriman's Elements of Mechanics............................12mo, 1 00
    Mechanics of Materials..................................8vo, 5 00
* Michie's Elements of Analytical Mechanics....................8vo, 4 00
Robinson's Principles of Mechanism............................8vo, 3 00
Sanborn's Mechanics Problems............................Large 12mo, 1 50
Schwamb and Merrill's Elements of Mechanism.................8vo, 3 00
Wood's Elements of Analytical Mechanics......................8vo, 3 00
    Principles of Elementary Mechanics.......................12mo, 1 25

MEDICAL.

* Abderhalden's Physiological Chemistry in Thirty Lectures. (Hall and Defren)
                                       8vo, 5 00
von Behring's Suppression of Tuberculosis. (Bolduan)............12mo, 1 00
* Bolduan's Immune Sera......................................12mo, 1 50
Bordet's Contribution to Immunity. (Gay). (In Preparation.)
Davenport's Statistical Methods with Special Reference to Biological Variations....................................16mo, mor. 1 50
Ehrlich's Collected Studies on Immunity. (Bolduan)..............8vo, 6 00
* Fischer's Physiology of Alimentation..................Large 12mo, cloth, 2 00
de Fursac's Manual of Psychiatry. (Rosanoff and Collins)....Large 12mo, 2 50
Hammarsten's Text-book on Physiological Chemistry. (Mandel).......8vo, 4 00
Jackson's Directions for Laboratory Work in Physiological Chemistry...8vo, 1 25
Lassar-Cohn's Practical Urinary Analysis. (Lorenz)..............12mo, 1 00
Mandel's Hand Book for the Bio-Chemical Laboratory.............12mo, 1 50
* Pauli's Physical Chemistry in the Service of Medicine. (Fischer)....12mo, 1 25
* Pozzi-Escot's Toxins and Venoms and their Antibodies. (Cohn).......12mo, 1 00
Rostoski's Serum Diagnosis. (Bolduan).........................12mo, 1 00
Ruddiman's Incompatibilities in Prescriptions.....................8vo, 2 00
    Whys in Pharmacy......................................12mo, 1 00

16

Salkowski's Physiological and Pathological Chemistry. (Orndorff).....8vo, 2 50
* Satterlee's Outlines of Human Embryology 12mo, 1 25
Smith's Lecture Notes on Chemistry for Dental Students...............8vo, 2 50
Steel's Treatise on the Diseases of the Dog..........................8vo, 3 50
* Whipple's Typhoid Fever.................................Large 12mo, 3 00
Woodhull's Notes on Military Hygiene 16mo, 1 50
*   Personal Hygiene.......................................12mo, 1 00
Worcester and Atkinson's Small Hospitals Establishment and Maintenance,
    and S ggestions for Hospital Architecture, with Plans for a Small
    Hospital 12mo, 1 25

METALLURGY.

Betts's Lead Refining by Electrolysis................................8vo, 4 00
Bolland's Encyclopedia of Founding and Dictionary of Foundry Terms Used
    in the Practice of Moulding 12mo, 3 00
        Iron Founder 12mo, 2 50
        "   "   Supplement 12mo, 2 50
Douglas's Untechnical Addresses on Technical Subjects...............12mo, 1 00
Goesel's Minerals and Metals: A Reference Book ....16mo, mor. 3 00
* Iles's Lead-smelting 12mo, 2 50
Keep's Cast Iron 8vo, 2 50
Le Chatelier's High-temperature Measurements. (Boudouard—Burgess) 12mo, 3 00
Metcalf's Steel. A Manual for Steel-users 12mo, 2 00
Miller's Cyanide Process 12mo, 1 00
Minet's Production of Aluminium and its Industrial Use. (Waldo) ...12mo, 2 50
Robine and Lenglen's Cyanide Industry. (Le Clerc) 8vo, 4 00
Ruer's Elements of Metallography. (Mathewson) (In Press.)
Smith's Materials of Machines....................................12mo, 1 00
Tate and Stone's Foundry Practice. (In Press.)
Thurston's Materials of Engineering. In Three Parts 8vo, 8 00
        Part I.   Non-metallic Materials of Engineering and Metallurgy ...8vo, 2 00
        Part II.   Iron and Steel...................................8vo, 3 50
        Part III.   A Treatise on Brasses, Bronzes, and Other Alloys and their
                    Constituents....................................8vo, 2 50
Ulke's Modern Electrolytic Copper Refining 8vo, 3 00
West's American Foundry Practice 12mo, 2 50
        Moulder's Text Book 12mo, 2 50
Wilson's Chlorination Process 12mo, 1 50
        Cyanide Processes......................................12mo, 1 50

MINERALOGY.

Barringer's Description of Minerals of Commercial Value.......Oblong, mor. 2 50
Boyd's Resources of Southwest Virginia.............................8vo, 3 00
Boyd's Map of Southwest Virginia. ....Pocket-book form. 2 00
* Browning's Introduction to the Rarer Elements......... .. 8vo, 1 50
Brush's Manual of Determinative Mineralogy. (Penfield).........8vo, 4 00
Butler's Pocket Hand-Book of Minerals.......................16mo, mor. 3 00
Chester's Catalogue of Minerals............................8vo, paper, 1 00
                                                                Cloth, 1 25
* Crane's Gold and Silver...8vo, 5 00
Dana's First Appendix to Dana's New "System of Mineralogy..". Large 8vo, 1 00
        Manual of Mineralogy and Petrography.......................12mo 2 00
        Minerals and How to Study Them 12mo, 1 50
        System of Mineralogy........................Large 8vo, half leather, 12 50
        Text-book of Mineralogy.....................................8vo, 4 00
Douglas's Untechnical Addresses on Technical Subjects.............12mo, 1 00
Eakle's Mineral Tables...8vo, 1 25
        Stone and Clay Products Used in Engineering. (In Preparation.)

Egleston's Catalogue of Minerals and Synonyms..........................8vo, 2 50

Goesel's Minerals and Metals: A Reference Book...............16mo, mor. 3 00

Groth's Introduction to Chemical Crystallography (Marshall)........12mo, 1 25

* Iddings's Rock Minerals 8vo, 5 00

Johannsen's Determination of Rock-forming Minerals in Thin Sections....8vo, 4 00

* Martin's Laboratory Guide to Qualitative Analysis with the Blowpipe.12mo, 60

Merrill's Non-metallic Minerals: Their Occurrence and Uses..........8vo, 4 00

    Stones for Building and Decoration............................8vo, 5 00

* Penfield's Notes on Determinative Mineralogy and Record of Mineral Tests.

                                          8vo, paper, 50

    Tables of Minerals, Including the Use of Minerals and Statistics of

        Domestic Production.......................................8vo, 1 00

* Pirsson's Rocks and Rock Minerals..............................12mo, 2 50

* Richards's Synopsis of Mineral Characters..........12mo, mor. 1 25

* Ries's Clays: Their Occurrence, Properties, and Uses..............8vo, 5 00

* Tillman's Text-book of Important Minerals and Rocks...............8vo, 2 00

MINING.

* Beard's Mine Gases and Explosions........................Large 12mo, 3 00

Boyd's Map of Southwest Virginia.....................Pocket-book form, 2 00

    Resources of Southwest Virginia.............................8vo, 3 00

* Crane's Gold and Silver 8vo, 5 00

Douglas's Untechnical Addresses on Technical Subjects..............12mo, 1 00

Eissler's Modern High Explosives. 8vo, 4 00

Goesel's Minerals and Metals: A Reference Book..............16mo, mor. 3 00

I lseng's Manual of Mining...8vo, 5 00

* Iles's Lead-smelting..12mo, 2 50

Miller's Cyanide Process...12mo, 1 00

O'Driscoll's Notes on the Treatment of Gold Ores. 8vo, 2 00

Peele's Compressed Air Plant for Mines 8vo, 3 00

Riemer's Shaft Sinking Under Difficult Conditions. (Corning and Peele)...8vo, 3 00

Robine and Lenglen's Cyanide Industry. (Le Clerc)..................8vo, 4 00

* Weaver's Military Explosives...8vo, 3 00

Wilson's Chlorination Process...12mo, 1 50

    Cyanide Processes...12mo, 1 50

    Hydraulic and Placer Mining. 2d edition, rewritten.............12mo, 2 50

    Treatise on Practical and Theoretical Mine Ventilation...........12mo, 1 25

SANITARY SCIENCE.

Association of State and National Food and Dairy Departments, Hartford Meeting,

       1906...8vo, 3 00

    Jamestown Meeting, 1907....................................8vo, 3 00

* Bashore's Outlines of Practical Sanitation........................12mo, 1 25

    Sanitation of a Country House...............................12mo, 1 00

    Sanitation of Recreation Camps and Parks...................12mo, 1 00

Folwell's Sewerage. (Designing, Construction, and Maintenance)........8vo, 3 00

    Water-supply Engineering....................................8vo, 4 00

Fowler's Sewage Works Analyses..12mo, 2 00

Fuertes's Water-filtration Works..12mo, 2 50

    Water and Public Health.....................................12mo, 1 50

Gerhard's Guide to Sanitary House-inspection...........................16mo, 1 00

*   Modern Baths and Bath Houses...............................8vo, 3 00

    Sanitation of Public Buildings................................12mo, 1 50

Hazen's Clean Water and How to Get It........................Large 12mo, 1 50

    Filtration of Public Water-supplies............................8vo, 3 00

Kinnicut, Winslow and Pratt's Purification of Sewage. (In Press.)

Leach's Inspection and Analysis of Food with Special Reference to State

       Control...8vo, 7 00

MISCELLANEOUS.

HEBREW AND CHALDEE TEXT-BOOKS.